Bushman of the Red Heart

Central Queensland
UNIVERSITY
PRESS

Bushman of the Red Heart

Central Australian Cameleer and Explorer Ben Nicker
1908–1941

Judy Robinson

An imprint of CQU Press

First published by
Outback Books
Central Queensland University Press
PO Box 1615, Rockhampton, Queensland 4700
Reprinted 1999

Robinson, Judy, 1931–.
Bushman of the red heart: Central Australian cameleer and explorer Ben Nicker 1908–1941.

ISBN 1 875998 70 5

1. Nicker, Benjamin Esmon. 2. Explorers — Australia, Central. 3. Cameleers — Australia, Central. 4. Australia, Central — History — 1905–1941. I. Title.

994.2

Design & Typesetting: Watson Ferguson & Co., Moorooka, Brisbane
Cover Design: Judy Robinson
Printed and Bound by Watson Ferguson & Co., Moorooka, Brisbane

Contents

Acknowledgements

This story is a compilation of memory, family records and tales told. I have cross-checked dates from sources too diverse to list. Michael Terry's books together with letters, journals, tape recordings, newspaper clippings and scrapbooks have been the source of much of my information.

Heartfelt thanks to Alan Low of Terrey Hills in NSW for his input, especially on military detail. To Ushma Scales, anthropologist of Alice Springs, for vetting Aboriginal content. To my family and friends for their encouragement and to my publisher, Professor David Myers, for his extraordinary patience.

Dedication

For Jane who fell in love with an Australian bush man.

1
Sam and Liz Nicker: Outback Pioneers
1901–1914

Ben's Mother, Liz Nicker

Elizabeth Elliott was the youngest daughter in a family of sixteen. Her parents, William and Anne-Jane were selectors and hoteliers of the Neve River and Blackall districts in outback Queensland. At the age of twenty one, Elizabeth married George Patrick Doolan and within the subsequent few years produced two daughters, Anne-Jane and Leah Clara. There are divided opinions over what happened next, but she left the marriage suddenly and found refuge with a sympathetic sister and brother-in-law out of town and beyond the wrath of her parents. It is suspected that George already had a wife in South Australia. Perhaps Elizabeth was the family 'bolter'. She was certainly the black sheep. Her family had very little to do with her, if anything, after that time.

Her sister's sheep property was small and wool prices were low. Not wanting to be more of a financial burden than necessary, Liz managed to find herself work as a shearer's cook on surrounding stations. She was forced to leave her infants in her sister's care for sometimes weeks at a stretch but this was a better solution than any other. She was able to proffer her meagre earnings and was sometimes able to bring back left-over meat and vegetables for the table. Every little bit helped.

In 1897 outback Australia was stumbling politically and economically through high hopes and empty pockets. British banks had withdrawn support for the Australian wool industry when the shearers' rebellion literally set the outback ablaze. When Australian banks crashed in repercussion and land companies went into liquidation, city streets overflowed with the hungry and homeless. State premiers were too ambitiously engaged in strategy and counter-strategy in their attempts to weld all the individual Australian states into federation. Poverty would hopefully sort itself out. It was a long battle and the young nation's economy endured a hardship almost beyond endurance.

Subsistence was nevertheless more manageable in rural areas and for many, never before had the prayer 'Give us this day our daily bread' been more heartfelt.

Heading home one evening riding side-saddle, Liz came upon a sheep struggling in a waterhole where it had come to drink. It was a dry season and the water had shrunk to little more than an unsavoury boggy puddle. There being no other soul in sight, she considered the situation. Had she not come along, the sheep would undoubtedly have died. That she had chanced upon it seemed to her to be a God-given opportunity. Ladies of the era never rode out alone. Liz was a rare exception but she usually carried a knife hidden somewhere about her garb for personal protection. Thus armed and forgiving herself for what she was about to do, she slaughtered the sheep and tied it across the front of her saddle. By that time night had overtaken her and she was able to hurry home under cover of darkness. Because she was only a slim little five foot two, the undertaking had required determined effort.

Visiting the area, a young solicitor called Banjo Patterson wrote 'Waltzing Matilda' in recognition of conditions of the time. Many men roamed the bush living off the land and forever walking towards probable work or rumours of gold hundreds of miles further inland.

Despite the times and her immediate situation, Liz had an almost over-abundance of personal dignity. She never discarded her corsets, hats, white cotton gloves, petticoats, stockings, or raised her hems, although patched they might be. She drove the buggy sitting rigidly erect while one hand held aloft her sunshade. The horses would have had to have bolted before she would discard her sunshade.

She inherited a genetic tendency turning her lovely hair white while she was in her twenties. While her contemporaries clung to their cumbersome long locks, which required time and effort to confine, Liz took the scissors to hers and her short white wash-and-wear hair-do set her apart.

For want of anything better, she massaged her hands with mutton-fat and sugar perfumed with lavender. She wore 'shirt-waisters' buttoned from throat to cuff above skirts sweeping to the ankle and a drift of lavender attended her every movement seeming to waft out of the dust underfoot.

She had 'what it took'.

Ben's Father, Sam Nicker

Samuel Foreman Nicker had been married and living with his first wife and their two small daughters in outback Queensland. They lived a gypsy lifestyle, moving with whatever work was available, contract horse-breaking, dam-sinking, fencing and shearing. It wasn't an ideal situation but ideal situations were impossible to find.

When one of the little girls developed a fever which proved serious, Sam saddled a horse and hurried away for medical help. He had a long way to go and on his eventual return with a doctor it was too late. Their grief was the final straw in the already troubled marriage and his wife and remaining daughter packed up and moved to Brisbane and the divorce court.

Sam sported a walrus moustache probably as much to compensate for the lack of hair on his head as for fashionable reasons. He was well aware life wasn't meant to be easy but it was his belief that results depended, whatever the problem might be, on your attitude. His was nonchalant.

He was a renowned horseman. He had ridden steeplechasing, followed the hounds and in his youth while still slim, had ridden as a gentleman jockey. Together with his mate, Harry (the Breaker) Morant, he had broken in horses for a living. He had a lifetime dislike of the term 'breaking in' and preferred to speak of it as 'educating' for which he had a natural knack.

He and Harry worked together with the legendary Jackie Howe as blade shearers and the people they met and tales they heard throughout the outback inspired in them a rivalry as poets. Their early efforts were published in *Smith's Weekly* and *The Bulletin* under thinly disguised pen-names and until Harry's regrettable demise in South Africa they kept up a constant correspondence. Before he sailed on that final trip Harry sent Sam a parody:

The minstrel boy to the war has gone,
In a Sydney bar you'll find him.
The stumpy mare his pack has on
And the Warrego lies behind him.

a verse is missing but it ends ...

And he'll drink one straight
For Sam his mate
And damn old Bobby Warner.

Bobby Warner was the local storekeeper to whom Harry owed money.

Sam was vital, optimistic and adventurous. Buried in his family's past was a colourful story of castles and counts, princes and bankers, generals and politicians, all woven into the fabric of the history of Germany and France. The name derived from a mythical river beast, half-horse and half-man, who was said to drag maidens down into the depths at night. Perhaps it was a story told to warn unsuspecting young ladies that river banks were dangerous places to be after dark.

Generations of Neckers, most of them Lutheran pastors, lived and died obscurely in Brandenburg and Pomerania. The genealogical march towards Sam included relatives involved in both the French Revolution and the American War of Independence, writers, professors, farmers and butchers. The usual mixed bag. Sam's father, also a Samuel Foreman Nicker, butcher, left his home on the Isle of Mann in 1853 and emigrated to Australia aboard the 'Caroline Chisholm'. He established his business at Eaglehawk near Bendigo, Victoria and on New Year's Day, 1860 married Margaret Houston Thompson, daughter of James Thompson, a Scottish professor of music. Young Sam was born in 1863, the second of six children.

Sam was hand shearing at Warbreccan Station near Jundah in November 1901 and was the union chairman when he and Liz met. We know nothing of their romance except that they fell in love and struck out seeking a new beginning. Obviously for both of them it was important to leave behind their former lives and start again beyond the boundaries of what they had individually experienced.

While they faced their future, there were regrets. Sam was moving beyond the possibility of seeing his surviving daughter ever again. For Liz, the struggle of separating her girls from her sister had resulted in a compromise. She allowed herself to be talked into leaving Leah, deemed too young to travel the distance where no proper roads existed. It was agreed that once settled somewhere, Liz would return for her.

Sam & Liz Travel Down the Cooper in 1902

Heading south by south west, they found large sheep properties where Sam was able to win employment, and waterholes strung like beads throughout the length of Cooper's Creek where they could fill their canteens, wash themselves and water their horses before moving on to good horse-feed.

It was never considered proper to camp on a waterhole because the presence of people might inhibit animals needing to drink. Neither was it advisable to use another man's campfire ashes in case he had disposed of strychnine. A number of men wandered the outback at that time, poisoning dingoes and scalping them. It was a precarious living but it was survival.

They were able to exchange their catches for cash whenever they found a police station and many were the tales told of deceits practised on scalps to double their earnings.

With Sam, Liz and Anne-Jane, all they owned was carefully stowed on board their baggage. It wasn't a large buggy and they didn't expect it would be an easy trip.

Their two horses were mere ponies but Sam had selected them carefully. He was, however, less cautious when along the way he noticed an excellent supply of honey in a high tree. Taking a bucket and a rope he scaled the trunk and while seeking to position himself was accosted by angry bees. So vicious was the attack that the victim was forced to retreat leaving his dignity among the branches with the rope and bucket. The equipment was a serious loss. Sugar was in short supply and for a taste of honey Sam was prepared for any effort but he would never again go anywhere near a bee hive.

When they arrived at Quorn, in the Flinders Ranges in 1902 Liz gave birth to a son, Samuel Claude, always known by his second name. Sam took a job with a local butcher and they stayed awhile but this wasn't their destination. The Centre was. They knew about the Arltunga gold rush and were headed in that direction with the hope of finding themselves a living and a pastoral property in newly opened country. When the baby was judged strong enough to travel, they headed north.

North to Arltunga Goldfields in 1903

Arriving at Marree, the travellers shook the dust from themselves and their belongings and wondered if this was the track they should be taking. Thirty five miles further north they came to a camel depot at Frome Creek. There were nearly three hundred camels and calves in hand and Sam was so interested that they spent a week with the community. It proved to be time well spent because most of these teams carried mail and goods from the railhead first at Marree then at Oodnadatta to all of the outback. Liz and Sam were able here to learn more of their destination and the road than they had previously known and met men who would ultimately become part of their background. They met and liked Charlie 'Bumblefoot' Kunoth, 'Pandy' Jack Gerard from Pandy Pandy Station, 'Scandalous' Jack Scanlon and Jackie 'the Rambler' among others men whose nicknames became part of outback folklore.

Further along the track between Frome Creek and Oodnadatta they came upon a rough sign nailed to a tree which warned,

Damn Arltunga, Damn the track
Damn it up and Damn it back.

Encouraged by the camel-men, 'As long as you keep a straight head on your shoulders, you won't have any problems,' they ignored the sign but they were to remember it on many occasions. Men died of thirst along this route and scurvy took its toll. In mining camps fever dispatched many victims and distance itself was the enemy as much as was summer heat and the lack of water.

The South Australian Government provided a string of wells for the movement of stock and travellers and the greatest care was required between them. Many of the wells had been sunk and equipped by a small team of men led by Ned Ryan and often referred to as the Ryan's Camel Well Sinking Party. A cumbersome name but a valiant effort. At a later date, specific 'whips' were designed and erected above the deepest wells where something more elaborate was required to more easily bring the water to the top. It was obvious that hauling up buckets by windlass was a difficult way to water a mob of thirsty travelling cattle and the 'whip wells' proved an invaluable step between the windlass and the windmill. To prevent indigenous animals falling in, each 'whip' was erected above a tall cairn of stonework, and a long race led out to a pulley. A horse, donkey or camel was harnessed to 'walk the race' while two huge buckets see-sawed alternatively up and down the well. No longer was an operator required to wind the windlass. Now all he had to do was grab a bucket and direct the water into a tank by way of a chute. A float regulated supply to the trough and as long as the float never jammed, the system was a good one. It was a system which played a very important role in the development of the outback at the turn of the century.

Women and children had seldom crossed this country and Liz and Sam were discerning in their progress north. They had to be.

It wasn't a faceless road for there were Overland Telegraph Stations and they were manned. It was possible to send a message, to stock up on stores, within reason, and even to bathe and wash clothes. The stations were situated a couple of hundred miles apart and it was a long road between them but travellers were few and were welcomed by the settlements. To pull a chair up to a table and dine in comfort was a luxury; walls enclosed them and there were comfortable beds for the night. It helped to hear of other people and places when their world had been reduced to a sometimes-barely-discernible dirt track.

Rising in the early morning Sam checked, repaired and replaced horseshoes, examined harness, adjusted load and greased axles. Every nut and bolt secured, he double-checked. Their very lives depended on meticulous attention to detail. As much as the travellers were eager to make a few more miles, so were thc settlement residents reluctant to part with their company. It could well be months before anyone else came along.

For some days they travelled in the company of the Turner brothers who shared one bicycle. They took it in turns and when the distance was judged enough, one would leave the laden bike for the other. Thus they passed by with a frequency guaranteed to keep company with each other and were grateful to share time and meals and sometimes a seat on the buggy with the Nicker family. Further along a packhorse mailman had lunch with them and brought news back from the goldfields and the Telegraph settlements along the way. It was a thousand-mile track they followed along the telegraph posts from one water-supply to the next, skirting rocky outcrops and sand-dunes where they could. They were travelling in the winter months and previous rains had freshened up the country through which they passed providing ample feed for their horses.

Creek banks sometimes held them up but they would make camp and settle down to shovel out a more accessible route. Creeks washed out with every rain and would continue to bedevil overland travellers for decades to come. The Depot Sand Hills tested them but they were fortunate because a shower of rain preceded them. It was still difficult progress. The dreaded Depot Sandhills roll away in east/west dunes. For twenty miles the travellers were forced to plough up and over, down and up again at right angles. Because the sand is ever-shifting they found no tracks to follow. They were frequently forced to unload and reload, to dig and corduroy and when once across after a three-day battle, they gave themselves a break to wash clothes and rearrange loading, bake bread and meat, rest the horses and overhaul the buggy. Ahead, worse awaited them. They had been warned.

A mountain range appeared in the distance and stretched away beyond imagination both east and west. They had rough sketch-maps with them and were therefore assured of gaps through which they could pass but from a distance the mountains seemed impenetrable, a barrier to the future they sought. They crossed wide plains where stands of desert oaks shrouded dark, cool oases of shadow. Wildflowers enchanted them and the scenery promised more than they had come to expect.

Their track led toward a ragged red line of hills and through a hewn pass marked on Sam's rough map as Devil's Gate. Its walls were etched with names of travellers who had preceded them. Devil's Gate was a man-made cutting hewn out of solid rock in 1896, through which 42 horses heaved a massive boiler to Arltunga. The South Australian Government, heeding the need for a Government battery cyanide works, had contracted Steve Adams to manage haulage. It proved an effort demanding endurance, skill and determination. There were days when horse-power had to be augmented with human ingenuity, crowbars and winches. Adams and his party achieved the near-impossible.

Steve hauled more easily the materials for the building of Central Australia's first hotel, 'The Cross Roads' at Arltunga and the roofing for the first few solid buildings. He was kept busy because while the mining field expanded, his was the only transport system there was and he had more than proved his capabilities.

In their footsteps, the Nicker family found a meagre track winding torturously through and along a creek bed strewn with boulders and shouldered each side by steep rockbound hills. Above them, occasional boulders were split open like eggshells revealing smooth white-lined caves which beckoned the wind and played mournful, abstract dirges. When you are the only people in a landscape every sound seems personally directed at you and the effect in these hills did not seem friendly. They were glad to see the last of it.

Flat plains and foothills alternated towards the mountain pass which led to their destination, the goldfields. The path climbed and twisted down into deep gullies and creek-beds, and wound around and steeply upwards to hilltops from where the views as much as the exertion took their breath away. They had frequently to block the wheels with rocks and logs and they found it simpler to walk beside their straining horses, urging them and soothing and protective of them. Slow and difficult as was this part of their journey they were buoyed by the knowledge that their destination was at hand and when they wound out of the foothills beyond, they could see in the distance, signs of habitation. It was Paddy's Rockhole, a little settlement preceding the mining fields. The place was called 'Annurra ntinga' meaning 'smelly water', but the native pronunciation sounded very like 'Arltunga' to white men's ears and the goldfield adopted the name.

Joseph Helle and Issac Smith discovered alluvial gold near the waterhole in April 1887. Prospects of fortunes to be made were heightened by rumours of rubies but when samples proved to be merely garnets, not everybody was disheartened. Gold or rubies, it didn't matter which. The outback was a vast untapped mystery.

Syndicates were formed around city conference tables by well-dressed men while hungry individuals set forth pushing wheelbarrows or riding horses ill-conditioned for the unforgiving distances, fierce temperatures and waterless wastes. Many simply walked and some walked into oblivion for their remains were seldom found and even today a shifting sandhill might leave uncovered a remnant of another questing soul.

The better-equipped and those who travelled in cooler winter months arrived to find conditions somewhat ruthless. The field wandered over a

A still-standing chimney at Arltunga goldfield.

Below: Mags in her eighties revisits the settlement.

Ben at work, aged seventeen.

The remains of Glen Maggie homestead, Ryan's Well.

Walers running free in Central Australia.

Eugene, Liz and Mags, and Ben (below). Both photographs taken in 1928.

Claude and Mags taken in 1917.

A Conservation Commission depiction of the Ryan's Well whip well.

Mags at the age of twenty one. 1931.

First race meeting at Barrow Creek in 1926. This photograph was taken by George Birchmore. He has written on the back 'All the white women between Alice Springs and Katherine at that time. L. to R. Phil Windell, Mrs Crook, Mrs Bohming, Elsie Bohning, Mrs Nicker and Margaret Nicker. Front row, Doreen and Kathleen Crook.

distance of 80 twisted miles and spread spasmodically to a width of 25 miles. The hills were scrubby and unattractive and clutched at the heat. It was stony and gave up its sparse treasures reluctantly. Water was in very short supply. In time, crushing mills, a warden's office, a police station and several stores alleviated some of the problems and later a bakery and pub made life much more bearable. New discoveries at nearby White Range encouraged the South Australian Government to install a ten-stamp battery and cyanide works and the official packhorse mail-service arrived more frequently.

When the gold output of Arltunga diminished, further discoveries at Winnecke's Depot in 1902 revived national interest.

The population of Arltunga, at its most productive, was around 400 men, but the field was never a rich one and the environment was forbidding. By 1916 South Australia was forced to close down its government battery and Arltunga gradually became another ghost town save for the few prospectors who stayed to haunt its outskirts.

The gold rush had, however, been a seed planted and from it grew the pastoral industry, the development of the township of Stuart, now known as Alice Springs, plans for the extension of the railway head which languished at Oodnadatta for almost forty years, roads, wells and a renewed interest in prospecting and surveying. It also opened for the first time in Australian history, a distinct overland reasoning instead of a sense of having to circumnavigate.

A white woman and children caused much astonishment when the Nicker family arrived in 1903. It had been a two-year journey. Their achievement of having conquered the track was applauded and furnished many a discussion around mining campfires in those initial weeks. Every assistance was offered as they settled in and a stone chimney went up and freestone walls which were clumsily built but sufficient to enclose their tent and living space. It was disheartening to find the field was on the wane but Sam found that rather than prospecting he could better employ himself carting. He bought a wagon from a disgruntled quitter and charged an appropriate fee for delivering water.

Liz started a garden on a near-bank of the Hale and when her melons, pumpkins, tomatoes, lettuce and onions ripened, she found a ready market.

Encouraged, she planted tobacco, cotton, cabbages, onions and potatoes. But the soil was poor and the work was hard and hot. They lived within a basin of low stony hills, glittering with quartz and mica particles. Heat seemed trapped. It was difficult for any breeze to find a path-way except along the little creek nearby, the only shaded respite within miles. She would take her infants down to play naked in the sand where she poured a little precious water to cool them down. She caught feral goats and milked them, providing milk and butter and sometimes small amounts of skimmed cream as well as meat and useful skins. From rendered fat and empty tins she fashioned slush lamps which burned wicks of saddle cloth or flannelette. A piece of moleskin wrapped around a stick held in place with sand was a better option but moleskin wasn't as easily found. Melted fat was poured over the device and allowed to set. The smell left much to be desired but any kind of light was desirable.

The family was practically self-sufficient except for staples like flour, tea, salt and sugar. These were available at the store throughout winter months but summer sometimes brought heavy rains and transport was difficult. There was a carrying service but in the wet it was a difficult undertaking. Mountainous areas and deep creeks cut across muddy plains bogging horses and wagons. It didn't rain often but there was never much food to be had on the goldfields at the best of times and careful management was required. Once established, Liz was usually able to provide extra meals for hungry miners who hadn't been as well organised with their supplies. The few pennies she was able to earn were a welcome addition to the family purse.

Using all their daylight hours as working time, most of Liz's customers for either meals or fresh vegetables arrived at night. Because there were no torches, they carried slush lamps to light their way or hurricane lamps when kerosene was available. Bobbing lamplight after dark imprinted an indelible delight in young Claude's memory.

Alice Springs/Stuart and the Overland Telegraph Line

The Overland Telegraph Line was open by December 1871 between Adelaide and Tennant Creek. Signals could carry only 300 miles, therefore repeater stations were built roughly 250 miles apart. The Alice Springs water hole was found to be an ideal site for a station and work was begun on the first building in November of the same year. It was originally built as a fort against Aboriginal attack but was never used for that purpose although the next station further north at Barrow Creek assuredly was. In the beginning Aboriginal attacks on white settlers were frequent. Some of the first pastoral leases taken up were abandoned for this very reason.

In 1874 a war party of Kaititja men attacked the Barrow Creek Telegraph Station. Because the evening was a hot one, the staff sat outside enjoying an evening breeze with Mounted Constable Gason. The war party had lain in wait on the flat-topped hills and crept down the slope under cover of approaching night. As the staff rushed for the cover of the building, one of the linesmen named Franks caught a fatal spear through his back and directly into his heart. Stapleton managed to get through the gate walls before receiving four spear wounds. His assistant Ernest Flint, and Constable Gason were both wounded but managed to reach their rifles within the safety of the building and frighten off their attackers. Flint then dragged himself to the office and sent off a morse code message to head office by relay through Alice Springs and Oodnadatta to Adelaide. As soon as news of the tragedy reached Charles Todd, he sent a carriage for Stapleton's wife and another for Doctor Gosse. The doctor, through a difficult morse-code process relayed instructions for treatment of the wounded but it was obvious that Stapleton could not recover. Aware of this, he asked to be lifted to the morse key and tapped out a last message to his wife and children, 'God bless you and the children.'

Police, trackers and settlers combed hundreds of square miles in retribution and shot every Aborigine they found, men, women and children. A watercourse about ten miles north of Ti Tree Well is called 'Skull Creek,' reminding us of the twin tragedy.

The Telegraph settlements were mainly self-sufficient, supplying their own beef and vegetables subsidized annually by supply wagons from Adelaide.

The isolation was crippling until the mining boom at Arltunga when men arrived in their hundreds, some of them ill and starved. The influx was beyond the Telegraph Station's capabilities and, as a result of a deputation to the South Australian Government, the plain two miles south of the settlement was gazetted as the township of Stuart in 1888.

Stuart, for some reason, was never an acceptable name as far as the locals were concerned. Government officials persisted in using it but the inland folk were used to calling the place Alice Springs or 'the Alice'. When the railway eventually arrived, the sign post designating it as Stuart was completely unacceptable and the Government apparently decided that the battle wasn't worth the effort.

On their first ninety-mile trip into town, the Nicker family approached it from the east, by way of Slip-Panel Gap and Undoolya. Their road led past Middle Ponds and across the Todd River to the Telegraph Station.

The postmaster, Thomas Bradshaw, was in residence with his wife Atlanta and their children, a governess and a staff of men whose jobs involved stockwork, gardening, upkeep of the telegraph lines, and blacksmithing as well as telegraphy and postal duties. It was a busy little community and an important one.

In 1901, four men driving back to the Telegraph Station from Undoolya capsized their buggy. A linesman, Harry Dixon, one of the passengers, suffered a badly broken thigh and it was the unlucky Mr Bradshaw who had the job of dealing with the injured man. Through relayed instructions by morse code from a doctor in Adelaide he performed a wonderfully successful piece of surgery notwithstanding the heavy doses of morphia he had mistakenly administered instead of anaesthetic.

The population of the little embryo township two miles to the south, numbered roughly twenty people. It was a much smaller community than the one at Arltunga. There was an iron-roofed store managed by George Wilkinson, and bearing a sign 'Wallis's store, agency and depot.' Opposite stood a thatch-roofed butcher's shop masquerading as a shed, neighboured by 'Charlie' Ah Hong's vegetable garden and 'eating house'. An hotel, The Stuart Arms, looked like an ordinary home, encircled by verandahs and hitching rails, under the management of a portly Mr Gunther and about to become the property of Charlie South. Running an hotel in such distant circumstances wasn't easy and the premises changed hands frequently. Mr South thought he might overcome the supply problem by building a brewery but that venture failed mainly due to the difficulty of carting sufficient water. Nobody particularly wanted the job.

Charlie Myers ventured into the Territory and took up the lease of Haast Bluff Station west of Alice Springs. A few years' hard work convinced him the Aborigines were eating better than he was so he sold the property and retired to the Alice. He built a home and separate saddlery shop where he was always busy and always available for a chat with any friends who happened along. Across the road was the Afghan settlement in the nature of combined loading depot and living accommodation attended by a makeshift mosque. The Afghan camel men were honoured members of the town community. They were the lifeline linking the Centre to everything they needed and keeping them in touch with the world outside. Annually, in celebration of the end of Ramadan, they gave a party to which the whole town was invited. Tarpaulins were spread with food and gifts and guests sat cross-legged enjoying more exotic fare than they were used to. When the town eventually produced children, they so much enjoyed the event that it became confused in their minds with Christmas.

A second store, Fred Ragget's, faced the hotel and two additional buildings, one empty and one providing shelter for Mr Wilkinson, were the only residences of any substance.

When Francis Birtles peddled through the Centre on his trusty bicycle in 1908 he estimated that at least a thousand camels browsed along the creek and spread amongst the saltbush.

A few transients and a caretaker camped near the government well. Gum trees shaded everyone and the creek produced little caches of soakage water. The town was entered by either an eastern approach from Arltunga, otherwise by a road along the Telegraph line bringing the traveller in from the west through Fenn's Gap. A cemetery clung to the dusty outskirts beside the western approach and goat yards circled Billy Goat Hill. A higher, self-contained hill rose between the township and the Telegraph Station and it was known as Kornerup's Bench until a later date when a memorial was built atop to commemorate the Centre's sons who gave their lives in the First World War and subsequently the Second.

Kornerup was a Telegraph Line worker who so much enjoyed the view from the top of the hill that he built a seat there so that he could more comfortably enjoy the spectacular sunsets. Not much is known about the man; he was one of many who laboured on the long and difficult task and then drifted into obscurity.

Cattle, horses and donkeys browsed, all keeping a comfortable distance from the camels armed with strangely jointed legs and able to kick in any direction. Stock-pads meandered from the well and crossed random wheel tracks grooved by horse-drawn vehicles and Aboriginal humpies were interspersed among low blue forests of saltbush and acacia.

A track led through the gap to the police station and it was possible to branch from there to pick up the telegraph track through Owen's Springs or to connect with the more frequented Arltunga road at Ooraminna Rockhole.

More than anywhere else in the outback, Alice Springs and the Stuart settlement had become the hub of a spider's web of bush-tracks leading outwards in every direction. The Centre was hesitating on the brink of a slow development.

A number of disappointed miners, perhaps having no home to which to return and unwilling to face the road back, drifted into the settlement or headed towards other prospecting possibilities and some of them left behind Aboriginal concubines and part-coloured children. Without any means of support or anywhere to live, it was a problem the townspeople solved in the only way possible. They banded together in 1914 and constructed a large tin dwelling behind the Stuart Arms Hotel. As it nearly approximated a bungalow-type residence it was soon referred to as 'the Bungalow'. Sometimes children were simply left there while their mothers went back to their former lives. This wasn't as heartless as it would appear. Their part-coloured children were not wanted tribally because they were considered 'wrong skin' and as such a threat to generations of careful reproduction. They were a contamination. There were instances of such children being ritually disposed of.

The few Aboriginal women who stayed were given employment sweeping floors, attending to hotel laundry and watering local gardens in return for food and clothing. Local cattlemen got into the habit of dropping off beef and a goat-herd provided milk and mutton. It was a satisfactory solution for all concerned until the children reached their early teens and their proximity to the hotel was deemed inappropriate. Nubile youngsters and boozing bachelors were an obvious recipe for disaster. In 1928 the 'Bungalow' was removed to an attractive out-of-town site at Jay Creek some miles west of the town, where it remained for many years.

The mining fraternity were multinational. They had wandered in from gold-seeking ventures in both the Americas, from the Yukon and from New Zealand. There were also refugees from an economically depressed Europe and from Russia, Holland, South Africa, the British Isles and Spain.

Most of them drifted further afield and grew old searching. Generally it was the married folk who settled the outback and took up pastoral properties. They 'invented' the outback.

An average pastoral lease measured roughly a thousand square miles. Of that area approximately one third was considered pretty good country. There was obviously plenty of waste land and a lot of distance between neighbours, and it demanded stout hearts, strong backs and high hopes. Above all, self-confidence.

Settler children grew up with a natural security in their environment and a unique relationship with the Aboriginal children they played with. For every settling family in the bush, it seemed an Aboriginal woman stepped into their lives to raise the children. They were very special people and much loved and respected, a bridge between two worlds. They were strict and humorous, patient and persistent. Born educators. Outback children were never aware until they grew up just how fortunate they were.

It was a harsh environment to outsiders. In March 1906 a prospecting party arrived at the Overland Telegraph Station by way of the Petermann Ranges. Their leader, F.R. George, was so physically debilitated that he died five days later. The party had been attacked by Loritja tribesmen in the previous December and T.W. Hall was speared in an eye while H. Fabian was wounded in the chest. Excessive summer heat and lack of water were serious conditions under which Mr George struggled to nurse his injured companions until they were able to resume their journey to Alice Springs. Deprivation, dysentery and anxiety generated the total collapse from which he was unable to recover.

Not only was it a difficult country to get into but it could be as daunting to leave. It was a long way to anywhere else. Suicide was a not uncommon event in the desert communities when men felt they had thrown their last dice. As the goldfields waned and conditions seemed more formidable, more and more men desperately considered their options.

Venereal disease circulated and there being then no known cure, 'death before dishonour' occasioned suicide in preference to an ignominious return home. The stress of summer's heat intensified disappointment and prompted fights and ill-considered departures.

Some miles to the north in the Hartz Range, Winnecke, another mining 'calico town', struggled in similar conditions; hot, dry, stony and unattractive. When typhoid swept through the community in 1903, a young bride gave birth only days before she herself succumbed to the fever. A young 18 year old packhorse mailman took the infant girl and delivered her to relatives at Oodnadatta. With all the canned Nestle's Milk the field could provide, he took the infant in a sugar-bag draped across his chest and tended it while performing his mailman duties across 400 miles of desert. Twice he was able to bathe the child at a homestead and bed it down properly to sleep. He had two riding horses and five packhorses which required hobbling out to feed at night and bringing in to saddle and load each morning. His hands were full but he delivered the baby safely at Oodnadatta without any ill effects save that it had grown used to sleeping on horseback.

Doctor H.W.Shanahan and a chemist, Mr Bishop, drove non-stop from Oodnadatta by buggy to help the stricken community. With only two little Timor ponies in harness, they broke all previous records and reached their destination in only seven days.

The Garden Settlement

Liz found herself pregnant again and when her time came a little earlier than expected, her second son, Eugene Foreman, was delivered by the field chemist/assayer. So delighted were the men of Arltunga that a white child had been born among them that little gifts of gold dust appeared and the chemist/midwife added a gold coin to the collection.

Months later the chemist died of suicide by arsenic poison and was buried in a nameless grave beside the Todd River, now a busy part of Alice Springs.

Liz had the coin set in a frame fashioned from the gold dust and suspended on a chain. She had the infant's initials and birth date engraved on the surround and added the initials of his midwife. This was probably the man's only memorial.

Liz wanted someplace healthier for her two growing children and they found what they sought further north along the Hale River. It was a pretty spot shaded by gums and well-watered. This house was a little more elaborately built of rough stone outer walls and a thatched roof of spinifex. There wasn't any glass for windows but Sam contrived shutters from kerosene tin boxes. There was a beaten ant-bed floor and food cupboards built out of crates. These were raised above the floor on jam-tins which were in turn set in tins of larger diameter filled with water, a bulwark against marauding ants and other crawling invaders.

No sooner settled in, Elizabeth again established a garden but this one was more productive and better able to supply fresh vegetables to both mining areas as it was midway between. The lack of fruit and vegetables was commonly known to cause 'Barcoo rot' often followed by 'Barcoo spews' and Liz's garden supplies were a defence against the problem.

Lacking any other form of entertainment, once night fell, the miners regularly indulged in gambling and thus their guns, horses and equipment frequently changed hands. During these exchanges Sam bought any unwanted available horses. He also trapped brumbies which had been released or lost and bred in the hills where they roamed free. They were easily caught when their thirst brought them in to soakages to drink. He built trap yards from rough timber and saplings. This was an ingenious maze through which animals wended their way to water but were then unable to find a way out. The original design was thought to have filtered into Australia from the American Indians.

When he had sufficient numbers, he trained them to either saddle or pack, and herded them north to Newcastle Waters, the Top End droving crossroads, where good horses were always in demand.

The Central Australian country, much of it stony limestone, was discovered to be excellent for horses. Native grasses were adequate and the topography developed good strong feet, sturdy legs and above average stamina. Because they were free-range and self-supporting, they were considered more intelligent than usual.

The Indian Army provided a ready market for horses at the turn of the century, during the years of British Rule. Thousands of mounts were required for the Boer War Cavalry, and then the Middle Eastern campaign of 1914–18. Australian Polo ponies were always in demand internationally. Many original pastoral properties were established primarily to supply overseas markets.

Selected through rigorous weight-hauling evaluation, heavy horses left over from the building of the Overland Telegraph Line were bred to lighter stock brought up from south and most particularly from the Hunter Valley in New South Wales. The sturdier mounts which resulted were trained to pull gun carriages and supply wagons. Lighter horses of the same breed became officers' mounts or were treasured locally as stockhorses. Known originally as New South Walers, the contraction 'Walers' became the more favoured label.

So few travelled the road in those days that Sam seldom met anybody on his trips north with horses and liked to tell the story in later years of having rolled his swag one morning and mislaid his wallet. On a subsequent trip some ten months later he found the wallet where he had left it. It was during these trips that he chanced upon the place he later named Glen Maggie, but he had first to earn the where-with-all to buy the lease.

Meanwhile their settlement on the Hale had become widely known as 'The Garden' and it flourished. Again Liz was pregnant but this time Sam was with her when she delivered another baby boy. Ben put in his appearance at dawn one grey, stormy morning and was as welcome as the rain in this semi-desert country. It was 1908.

Billy and Clara walked into their lives. Liz had her hands full and was grateful to have help. They were an Arunda Aboriginal couple who for cultural reasons, (wrong skin), were irrevocably severed from their country and their tribe. Subsistence decreed that wandering desert communities were small in numbers. To protect themselves from social misalliances and inbreeding, an inflexible behavioural system governed relationships defining boundaries and governing sexual liaisons. A skin name was a group name within the tribe and a skin system is too complex to define within a few brief paragraphs. Any infringement was met with severe punishment, including a spear through the leg, banishment or even death by 'singing.' Preventative measures were learned customs. For instance, a girl-child walked with her eyes downcast in the presence of men. 'Can't look longa that one.' She was 'given' from birth to an older man of the correct 'skin' and handed over to the man's older wife at puberty. She fetched and carried, child-minded and was taught the skills she would need to take her rightful place in her community. She knew herself safeguarded and in later years she in turn would be waited upon and respected.

There were many among the white settlers who had high respect for the Aboriginal skin system.

Billy and Clara had managed to run away but they were aware that without protection, retribution would follow. Because any children of their taboo relationship would reflect their crime, they used herbal knowledge to prevent conception and extended to the Nicker children a kind of dignified foster-parentage. They shared their Arunda language and the dialects of surrounding tribal areas with the young Nickers. They gathered bush tucker, told their Dreamtime stories and passed on their tracking skills. In exchange they found a permanent home in the affections of the family and the children grew up balanced between two cultures.

Sam was now more often away, assured that Liz and the family were aided by Billy and Clara, who were, in turn, protected by Liz. She was never a good shot with the rifle but she was confident that all she would ever need to do was aim it in the general direction of any human threat. Because she was never threatened she never had cause to justify her theory. Her only real worry was snakes. There were death adders and king browns and unidentifiable little black snakes and she sometimes felt she had developed a permanent head-down stoop due to her watchfulness. She shook out bedclothes nightly and double-checked shoes. She swept inside and had Clara sweep around their home so that any snake tracks were followed up and the snake dealt with before her children were allowed out of bed in the morning.

There is a well-known tale of a policeman who had the same anxiety about snakes. He squatted down to place his quart-pot on the campfire one night while out on patrol. Feeling a sudden sting on his right buttock he leapt to his feet, dropped his trousers and insisted his Aboriginal offsider cut and suck out the poison. The poor man wasn't allowed to waste time with whatever it was he was trying to say until after the job was done. Only then did he have permission to remark, 'That was no snake, boss. You sat on your spurs.'

Sam busied himself taking any job which paid actual cash, a hard-found commodity in Central Australia. Because there were no banks the barter system was commonly used and 'shin plasters' enjoyed a brief viability. They were hand-drawn cheques in lieu of I.O.U's and bore a specified amount but the unscrupulous were inclined to bake them so that reposing in either pocket or wallet they crumbled. They were not a popular idea.

With his sights set on Glen Maggie, Sam began buying a few odd sheep here and there, building up his stock against the day. He was never a cattle man. He was from sheep country and considered a 'gun shearer', in other words a top man with blades. He knew sheep and felt certain that they would do well in this new part of the country. Like a lot of other people he was eventually disappointed but like everybody else he had to prove it for himself. Had wool markets been much closer to Central Australia, the many other problems might have been overcome.

When Ben was two years old, Liz packed the buggy and took Anne-Jane and the little boys with her to the settlement growing up around the Todd River a short way south of the Telegraph Station. She left Billy and Clara to mind the garden and tend the animals.

She had mail and stores to collect and although well along in another pregnancy did not expect the buggy trip would hurry things along quite as quickly as happened. Margaret was born in a tent hurriedly erected beside the buggy beneath gum trees near the government well. It was the eleventh of June, 1910, a night made memorable as the earth passed through the tail of Halley's Comet. The comet that year brilliantly lit the heavens and to many in the outback without access to newspapers, it was a startling and even frightening event. There were reports of a glowing night made more dramatic by occasional drifts of microscopic particles of light. One most unimaginative bushman was heard to suggest either he had drunk too much or had stumbled into fairyland. Another was camped with cattle and told a tale of a silence 'as if everything had forgotten to breathe.' A stockman riding night-watch was so enthralled that he found he had ridden in a straight line instead of around the herd and temporarily lost himself.

The Pioneering Hayes Family

Anne-Jane had just turned sixteen when Margaret was born. She had always been a pretty child and her developing charms had recently become a topic of interest throughout the Centre. The outback's young bachelors eagerly presented themselves but despite her youth, she had already chosen Ted Hayes, a tall young man offering his home at Mt Burrell, a property south of Alice and acquired by his family. He was the eldest son of a pioneering pastoral family who had entered the area as dam sinkers and stayed to make the Centre their home. Previously, as with so many others, theirs had been a gypsy lifestyle, going where the work was, fencing, carting, ploughing and scooping, until entering into a contract with Sir Thomas Elder to work on his property, Mt Burrell, about a hundred miles south of Alice Springs. William Hayes was employed to do all the fencing and dam sinking. He also carried telegraph poles for the Overland Telegraph and erected fencing on Owens Springs until his family, working with him, prevailed upon him to settle down.

They bought Deep Well, a small property running only 162 head of cattle and the family, working as a team and sometimes against all odds, gradually built up their pastoral empire, adding Mt Burrell, Undoolya, Maryvale, Simpson's Gap and Owens Springs. Delivering stock for sale to a property on the Diamantina River the family bought a flock of 1000 head of sheep. The menfolk had to return and left Mary, (Mrs Hayes) and her two daughters to overland the mob the 350 miles home. They ran into big rains and the sheep got frequently bogged but they pulled them out and slogged on. Mary lost her boots in a bog and finished the trek barefoot. It was all a misspent effort because sheep never proved a successful enterprise in the arid Centre. Burrs, dingoes and hungry Aborigines as well as summer heat, undependable water supplies and distance to wool-markets proved their undoing. They replaced sheep with goats and concentrated on cattle and horses and their properties thrived.

Everyone depended on their goat flocks. Without refrigeration, it was a less complicated matter to use the smaller carcase especially in summer months. Goat milk, cream, cheese and butter were dependable staples, and the skins were frequently softened and used for mats, seat covers, slippers and sometimes saddle-bags.

Alice Springs' First Wedding

The wedding of Ted Hayes to Anne Jane Nicker was the first one held in Stuart. The service was performed by the local police officer, he being then the only settler qualified, and attended by everyone able to get there. In time, the couple became scions of the district and Anne-Jane a recipient of the Order of the British Empire for services to the community.

In view of the fact that the family now travelled more frequently into town, Sam bought the slab cottage vacated by Sandy Myrtle. Known as Myrtle Villa, it represented a base a lot more comfortable than any they had previously enjoyed. As land was not expensive in those early days, he was also able to buy town blocks for his children evincing his faith in the future of the little settlement.

Meanwhile life was becoming more sociable. As the population expanded, the odd bush race meeting settled into a definite annual event and while ladies sewed by hurricane lamps and ordered multi-coloured hat veiling from southern emporiums, the menfolk polished up their buggies and readied their horses.

Stockmen arriving earlier at the racetrack bringing horses and setting up temporary yards, made camp at the base of a hill south of Heavitree Gap. Around the evening camp-fire they indulged in so much skiting and yarn-telling that the hill acquired the name, Mount Blatherskite. A square of canvas was pegged in preparation for dancing under the stars and children slept tucked up in swags beyond the fringe of adult activities. Musical instruments were produced from out of saddle bags, usually a violin, a banjo, a concertina and the odd mouth-organ or two. The instrument had to be sufficiently small to tuck carefully away with more important necessities. Popular music tended to be Irish and ran the whole gamut between nostalgia and lively dancing jigs. 'Take me home again, Kathleen' and 'Danny Boy' headed the list of favourites and everybody sang along.

Gossip and horses were exchanged and newcomers introduced. Women and children caught up with each other while their menfolk discussed market prices, seasons, droving trips and branding. As inebriation advanced, provocative bets were placed against each other's horses and/or riders. Cattle stations were sometimes wagered and there was once a splendid racehorse exchanged for someone's wife.

When finally the meeting ended, spontaneous buggy races would terrify the wives and exhilarate the children. While husbands challenged each other, ladies clung desperately to their sunshades and hats, streaming swathes of cherished veiling behind them. 'But last year you promised!' While children screamed, 'Faster, Dad, faster! They're gaining on us!'

There were also gatherings on the lawns at the Telegraph Station. Despite a ratio of ten men to one lady, dances were popular but even more delightful were the moonlit buggy rides back to town.

The outback was patrolled by a preacher who travelled by camel and was known throughout the length of his route north to Tennant Creek, as 'The Man from Oodnadatta'. His name was Bruce Plowman and he always came bearing gifts of magazines and sweets and bringing news of a wider, more sophisticated world. His parish extended over an area of 160 thousand square miles, 39 thousand square miles larger than the combined area of England Scotland and Ireland. The total white population numbered about four hundred compared with the many million in the British Isles. To cover his area annually he travelled 25 hundred miles. He christened the children and held small services when his offer was acceptable. He was always and everywhere a welcome guest, a different face from a different place and a connection with friends along the track.

When Liz insisted the time had come to baptise her two youngest, they were discovered to have gone bush but were, as ever, easily discovered by an alert Billy. They were rebelling against having water poured over them and being sworn at, but the deed was performed despite their evident antagonism and another outback christening tale was added to the lengthening list.

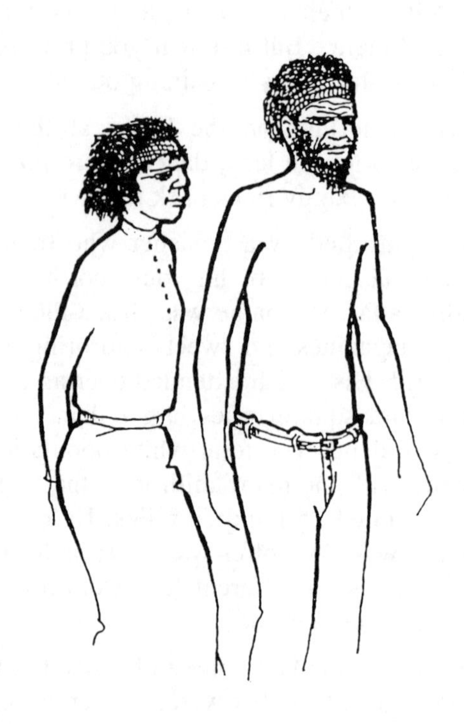

2
Ryan's Well and Glen Maggie
1914–1930

The Collision

Sam's dream became reality.

He was at last able to buy the Ryan's Well lease at the cost of a halfpenny per square mile. Commemorating both his mother and his youngest daughter, he called it Glen Maggie which gave rise to a lot of confusion since Ryan's Well and the eventual Glen Maggie homestead peered at each other across the road.

In August, 1914 the family packed up and headed the 78 miles north towards the home they had come so far to possess. While they were camped on the Burt Plain on their way, Ted Dixon, travelling from Alice Springs, caught up with them and broke the news that war had been declared in Europe.

Although the war was very distant they thought of themselves as living in an outpost of the British Empire and misgivings undermined their confidence. Even in their isolation they now had more to think about than establishing a cattle station.

Ben sat on an upturned basin among bedding, boxes of foodstuffs, crates of hobble-chains and neatly stowed and tightly-tied pieces of household furniture.

His father sang a Scottish air though the words were sometimes lost in the medley of grumbling iron wheels, slap of leather harness, an intermittent stumbling of horses hooves and an interjection now and then from the groaning timber parts of their overloaded wagon.

They had travelled north from the little settlement at the Alice Springs Telegraph Station and the trip was beginning to lose its charm.

He was six years old. A squarely-built boy, his hair was rust-coloured normally, but now he presented a dustier spectacle, very like a sepia photograph, saved only by the startling contrast of bright blue-grey eyes. They were humorous, reflective eyes but now they were sleepy and their lids were heavy. His interest in the track and the bush through which they travelled had kept him alert. Lizards, kangaroos and a snake, startled out of its loops to speed arrow-straight into undergrowth; a few dingoes padding inquisitively along keeping pace with them.

Yesterday they had trailed across a spinifex plain, relieved by sparse grey shrubbery and this morning everything had changed. They'd wound across a creek-bed in a gap in the Hann Range where pine trees dotted the hillsides and gum trees nodded in the early morning breeze. Bloodwoods harboured flights of brilliantly green budgerigars and screeching cockatoos.

Past the gap, they came into a wide, shallow valley where shadows dappled their track through softer grasses, herbage and mulga. The fierce spinifex plain behind them was restrained from infringing on the valley by the stolid demarcation of the Hann Hills. Ben loved the excitement of the bush.

A collision of sounds, an explosion of motion! The boy's dreaming body was launched from the pitching wagon and landed dangerously close to bolting horse's hooves. In an involuntary instinct he balled his limbs, ducked his head and rolled aside as the wagon-wheels missed him by inches.

What had gone wrong? Where the scrub had given way to level grassland and while the boy dozed, the family wagon drew alongside the buggy driven by his mother. While his parents spoke, their minds not entirely on their road, a rock in their path lurched the wagon forward against the buggy, locking wheels. The wagon wheel tore apart and the upturned dish beneath Ben shot overboard taking the boy with it.

Disentangling their frightened horses from tilting shafts, the two drivers ran back through whirling dust. They were halted in their panic by the sight of Ben's dishevelled figure struggling upright and announcing, 'Why don't you kill a kid outright, not just frighten him to death!'

Sam and Liz sagged against each other in relief surveying the child and the crippled wagon. There was a long silence while they collected their wits.

Finally Sam spoke, 'Well, my love, that wheel won't be going any further. It will take us a year to get a replacement so we'll camp here tonight and think on it.'

Pulling themselves together, they moved each to an appointed task, Ben to assist his father with unhitching and hobbling out their horses; Liz and the child Margaret, affectionately known as Mags, busying themselves gathering wood, lighting a fire and preparing a meal.

Last night they had camped for the first time within the boundary of their own property. The property consisted of about five hundred square miles surrounding Ryan's Well. It had been sunk and equipped about twenty five years previously as part of the Overland Stock route between Adelaide and Darwin and stood in the absolute heart of the continent.

There weren't any walls to surround them but they were home.

Their ultimate destination, the homestead site, was a further twelve miles away but here they were, within view of the well itself. There was sufficient water for their following stock although it was brackish and too unpleasant for human tastes.

Among the mulga a smooth red clearing of earth, faintly scored by lizard tracks, presented itself as a sensible camp site.

Before the following stock could catch, up a temporary night-yard had to be prepared. There were four hundred sheep, some goats and a small, mixed mob of cattle. They were shepherded along by Billy and Clara and Ben's two older brothers, Claude and Eugene, with fourteen riding horses and two pack mares with foals.

While Sam chopped, Liz and the children dragged up branches and helped stack them in a large circle. It was cool weather and this was a job they had become accustomed to. They had much yet to do when they heard the first faint sounds approaching. But first the animals would pass them by and water at the well. They would then be turned out to feed until sundown before they were yarded. There was plenty of time for explanations, questions and answers. Time for a meal in relays. Time for a gathering of resources.

The sun set that night through a powdering of dust, sprinkled liberally over grass, leaves, animals and people. It was not enough to notice and certainly hindered none of the evening activities. There was a discordant clunking of neck-bells and jingling of hobble-chains as their horses browsed beyond the perimeter of light radiating from the campfire. Shuffling noises, a few complaints voiced by the sheep and goats now settled in their temporary yard, all these sounds unregistered by anyone because they told the comforting tale that all was well.

A night owl whoo-whoo-ed enquiringly at all the unaccustomed activity. Rarely had there been such movement, so many people, animals and sounds within his knowledge. He settled himself on a branch near the trunk of a mulga tree and absorbed these new sights, swivelling his head now and then towards new sounds.

The fire's glow mesmerised him.

So taken up with curiosity was he that he forgot to hunt and stayed watching until the first faint touch of morning woke the camp and frightened him away.

Sam boiled the billy and enjoyed his first pannikin of tea for the day, sitting peacefully and drawing on his pipe. There was no rush to move out early and he took his time.

It was a new day and perfect August weather, crisp and cool with a gentle early breeze flirting with the campfire. A good day for ironing out the wrinkles in the wake of yesterday's calamity.

When the animals had been attended to and breakfast cleared away, the family walked the mile towards the well armed with soap and towels. The kids hurried on ahead while the adults strolled at a more leisurely pace examining their surroundings. They wandered up a gently rising stony slope and looked out in every direction at a pleasing view. It was a shallow valley and when Sam had seen it for the first time it had been shortly after rain, bursting with new green growth reminding him of pictures he had seen of Scotland and of his maternal Scottish heritage.

Could we move camp up here? It was a thought they considered carefully. They had to be a distance from the well to leave it free for watering stock, both their own and travelling herds. From this point they could themselves be a little more prominent and see further afield. They would have to cart their drinking water from the rockhole twelve miles away, negotiable by buggy until the wagon was back in use, but that was an eventuality wherever they made their temporary home. This wasn't going to be permanent. The ground where they had camped last night was lower-lying and would turn boggy when it rained. Taking everything into consideration, they concluded this was a better site.

Clean and changed, they were now ready for action. It was time to bring up the buggy, unloading it and returning again and again until the wagon stood tipped and abandoned where it had crashed.

The family worked as a team in relays while the younger two, Ben and Mags begged rides each way and pottered among the bundles and boxes.

By nightfall they were comfortably established in a temporary fashion on the ridge.

Both Claude and Gene were capable lads. Although only ten and eight respectively they could drive the buggy and harness up and liked to think of themselves as men. What they lacked in size they made up for in enthusiasm. At six Ben had yet to develop staying power. He was more a hindrance than a help.

As time passed the little camp on the ridge took on a more permanent air.

Ben and the Afghan Cameleers

The wagon wheel proved, as Sam had forecast, to be beyond repair. He had no access to a forge nor the metals required for the job and therefore an order was sent by way of a linesman attached to The Overland Telegraph. It was indeed almost a year before a replacement wheel arrived. It came overland roped to the side of a camel and balanced by a crate of goods headed further north.

During the years of the First World War, wolfram was an important adjunct to the war effort and Wauchope and Hatches Creek were two rich fields. Camel teamsters were employed from Marree and Oodnadatta carting supplies north and back-loading ore from the mines.

Although a few cameleers actually came from Afghanistan, others were from Rajasthan, Baluchistan and nearby areas. As the industry developed in Australia, more handlers arrived from Egypt, Persia and Turkey and they too came under the blanket term, Afghans. They came mainly from ancient desert landscapes and were comfortable with the emptiness of the Australian inland. They came in the flowing robes of their homelands but adapted to local wear in every aspect except that they never discarded their turbans and neither did they ever mislay the spirituality inbred over many centuries.

When the camel trains rested at Glen Maggie mending packs and splicing ropes, Ben attached himself to them. He was familiar with them. Usually, on this run were Sadiq, Aktar Mahommet and Ali Mahommed. Each man travelled with his Aboriginal or part-Aboriginal consort and their children and two or three Aboriginal helpers. They were not permitted to bring wives or family into the country from their homelands and had generally arrived as single men. The few deprived husbands and fathers among them sent large parts of their earnings home and some took Australian consorts and raised their second families with loving care and attention.

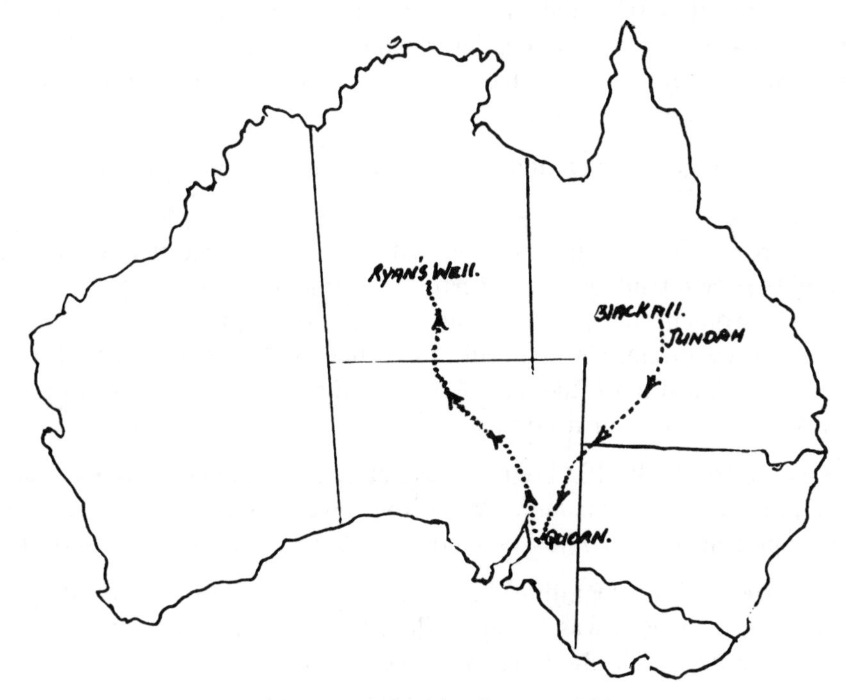

The route taken by Sam and Liz.

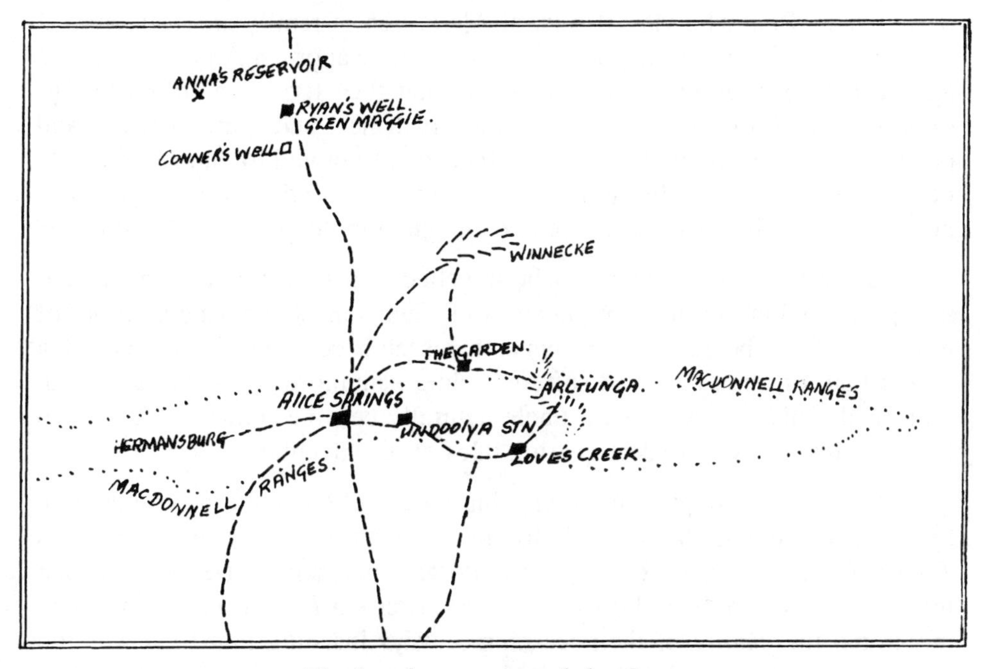

The local area around the Centre.

Wherever they travelled, their prayer mats accompanied them and left strange indentations across the deserts they traversed, for whatever their circumstances they never neglected to face Mecca in contemplation and communion with Allah.

They were kindly men and everyone watched for their arrivals.

They had special boxes which opened like cupboards and displayed small shops of needles and cottons, safety pins and buttons. Veiling and laces, ribbons and baubles, the cheerful, feminine things to delight the hearts of isolated women and their growing daughters and sweets to brighten the tastebuds of everyone. Children all along their route looked forward to the brief companionship of the camel-team children and relationships were forged which lasted their lifetimes.

On one trip Sadiq left behind a camel calf. Orphaned, it was raised on goat's milk back at Oodnadatta. When it was weaned, Sadiq considered Ben would appreciate it. An enthusiastic boy couldn't wait for his promised camel.

He dreamed of handling it, training it and riding about the bush exploring. In time Ben was to do all that. But while he was still a child he had things to do which filled his days, and lessons to learn.

His father was a wondrous story teller, tales which fed his imagination and fuelled his dreams. Sam, so isolated from his own background, liked to tell the children of progenitors they would never know. He spoke sometimes of an uncle of his who visited Australia in search of family and following at the same time rumours of gold. En route, at a seaport in America, he was approached by a group to join them in what they termed a 'filibustering' expedition to Mexico. They called him 'Britisher' and were friendly and persuasive but somehow he suspected their plans pointed to gun running. He politely refused to involve himself and when reaching Australia read in a newspaper that they had been captured and hung in the plaza in Mexico City.

Part of his luggage wherever he travelled was a dolly pot, a small cast-iron pestle and mortar used by prospectors. Sam's most enduring memory of him was of the beard he wore beneath his chin covering the effects of a severe bout of quinsy he had suffered while panning on a North American gold field. This particular great-uncle naturally became somewhat of a hero for Ben and he never tired of hearing tales of his adventures.

With every infrequent mail, arriving by camel team and now and again by travellers thoughtful enough to deliver mail from the Alice Springs Telegraph Station, came books and children's magazines. Determined that living as and where they did should be no deterrent to their education, Sam monitored the quality of their learning, always insisting that time was too precious to waste. Because the older boys were also his major work force, it was Mags and Ben who 'learned to learn.' They did their sums and studied history, English and geography. They recited lengthy stanzas of poetry, were taught Greek mythology and very basic astronomy. When a passer-by could be prevailed upon to stay awhile and teach, it was a welcome bonus adding a different flavour to their studies. The men who wandered the outback were well-read. The books they carried in their saddle-bags were the classics. There wasn't any room for anything which couldn't be read and re-read a dozen times and then swapped. Dog-eared, margins often scribbled on with the acid comments of previous owners, they were kept like icons, carefully wrapped in a square of saddle cloth or calico and sometimes smelling of the smoke of the many campfires which had lit their reading.

When they met, these travellers would sit hunched for many hours, arguing and discussing a passage or a book, while tit-bits of national and local news barely registered. Able to recite whole chapters of classical literature, they were in some instances unable to name Australia's current Prime Minister. In this distant outback they were too far removed to keep up with current affairs.

Robert Stott, Transplanted Scot

It had been Sam who was instrumental in bringing the first teacher, Ida Stanley, to the Centre, but when it happened his own children had the benefit of only six months before the family moved away to settle and establish their own new property.

A year previously, in 1913, Sam had fist-fought an irascible Aborigine who threatened dreadful consequences unless Billy and Clara were handed over for ritual punishment. Sam suffered a damaged eye which would not heal despite Liz's nursing skills. It was a thousand miles in any direction to reach qualified medical help but the simplest route was south. Sam was duly bundled on horseback and taken by Billy all the way to Oodnadatta with only a bottle of whisky against the pain. En route his eyeball burst.

At Oodnadatta they had a long wait for the train and when it came, Sam had of necessity to leave Billy with the horses to await his return. It was not an easy sojourn for Billy. This was way out of his own country and into vastly different tribal customs. He kept an anxious eye on their horses, hopefully watched the fortnightly arrival of every train, avoided more than the most necessary human contact and remained in almost constant view of the store. Sam had asked the storekeeper, a friend, to make sure Billy was cared for and supplied and he filled his idle hours with wood-chopping, sweeping, and happily turning his hand to whatever needed doing. As time passed, the storekeeper comforted him and assured him that Sam had not been swallowed up in the unknown but would most assuredly return when he could.

Meanwhile in Adelaide, while the socket healed and he was allowed some freedom as a walking patient, Sam spent much of his time listening to debates in Parliament House on North Terrace, a short stroll from his hospital ward. It was thus that he petitioned for a teacher in the Centre, backed up by correspondence from the resident police officer back in Alice, Sergeant Stott. His personality and qualifications suited him admirably for the position. For the years he served, he *was* Alice Springs. Robert Stott, the resident 'everything' in Central Australia also had a large family and their education to consider. It was Sam's initiative and persistence coupled with his association with the then Minister for the Interior, Patrick MacMahon Glynn, which finally upgraded Alice Springs to a very distant outpost of the South Australian Education Department.

Robert Stott was a transplanted Scot. When he arrived in the tiny settlement of Stuart with his wife and family, he soon found himself not only the police officer but also the Protector of Aborigines, the Stock Inspector, the local dentist, Coroner and Registrar of Births Deaths and Marriages. He was inspector and unofficial acting governor. The many thousands of square miles making up his district were in immense contrast to his native Scotland but he was an adaptable man, a reasonable and able officer, liked and respected.

Many were the tall tales told of Robert Stott.

Preparing her class for the impending visit of the Governor General so that they would be suitably impressed, their teacher carefully explained the hierarchy before questioning them.... 'and who is the Governor General answerable to?'......... 'and who is the King answerable to?' It came as no surprise when the class answered in unison, 'Sergeant Stott!'

To fill the card on annual sports days in town and make sure everybody was included in the festivities, there was usually a foot-race for Aboriginal ladies. Sergeant Stott discovered that one was more fleet of foot than most and bribed her with a new dress to win. However, word leaked out and some of the locals offered her two new dresses to lose. The bookies did a brisk trade and when the Sergeant discovered the double deal, his wrath was such that the perpetrators left town for a fortnight.

Fitted with a glass eye and armed with the promise of immediate attention from the South Australian Education Department, Sam eventually stepped off the train at Oodnadatta, to the relief of Billy who had believed himself forever stranded. The eye was for awhile a problem and difficult to manage but he learned to use it to good effect when faced with any difficult situation. Then he would delight in slipping it from its mooring and holding it forth with the pronouncement: 'This one debil eye. Watch you-fella all the time.'

Liz as District Nurse

Because there were few white women in the Australian inland at that time, Liz's good common-sense and self-taught nursing skills were frequently called upon. She was often away from home delivering babies or tending the sick. Sometimes the message would arrive by 'yabber stick', a letter carried in a cleft stick by Aboriginal courier. She would pack her bags and go alone by buggy sometimes distances of a hundred miles or so. If it was possible, one of her sons would accompany her, leaving Clara to attend to any emergencies at home. Sometimes her absences were brief but there were occasions when she was required for weeks at a time.

She never lost a patient but was once called too late to save a prematurely presented infant and could only concentrate all her efforts on the mother. A message had been sent by an unreliable source and should it have arrived even twelve hours earlier she had no doubt that she would have also saved the baby's life. When the message did reach her, just on sundown, she recognised the urgency and harnessing two horses, travelled the sixty miles without pause. Her road led over rough mountainous tracks which would have been difficult enough in daylight. She prayed for her patients and she prayed that in her haste, her horses would bear up. She didn't dare consider the consequences should a misstep take them into a ravine or tip them into a gully.

Today a stone carved by renowned sculptor, William Ricketts, commemorates Liz's nursing skills. Titled 'The Helping Hands', it can be seen at Pitchi Ritchi just south of Heavitree Gap in Alice Springs.

In his book, *The Man from Oodnadatta*, R.B. Plowman writes,

> In the annals of our race there have been recorded from time to time the names and noble doings of great women. In the lonely places of the vast Australian continent there are women whose names and deeds are worthy of record in these annals. One is Mrs Sam Nicker.
>
> *The Man from Oodnadatta,*
> R.B. Plowman, Angus and Robertson Ltd, 1933.

Wells and Waterholes

The family dug wells but found water in short supply. Well-digging was energy- and time-consuming. As they deepened, they were required to chop, square and cart timber with which to shore the sides. When the hole extended beyond light and air, they erected a slope of calico or canvas above the hole thus deflecting sunlight and any breeze to the worker below. Although the surface pretended evidence of underground water and they persevered to depths of up to two hundred feet, their efforts proved fruitless.

Deep digging was attended by danger. In 1903 Johannes Kadow, a mere lad, died from suffocation caused by foul air at the bottom of a well. He had been in Alice for a bit of a break with his brother. When he returned to work at Temple Bar Well, and was lowered into the depths, such thick, foul air had settled in the bottom that the young man quickly lost consciousness.

Sam extended Glen Maggie boundaries, adding first the Conner's Well block encompassing an additional 500 square miles and then Anna's Reservoir, bringing the total area to 1,500 square miles. They dug more unsuccessful wells and resorted meanwhile to sending their flocks to camp on enduring rockholes. Billy and Clara, accompanied often by Ben and Mags built semi-permanent accommodation at these water holes and whiled away their hours and sometimes months teaching the children bush survival and hunting.

Anna's Reservoir was discovered by John MacDouall Stuart and provided a key depot-point on his epic trips across the continent. It is protected on its northern side by a rock wall forming a natural dam, and great flat rocks have slipped into protective angles, cupping the waterhole. Small gum trees shade the approach along the creek and rock wallabies pursue their time-worn paths among surrounding rocks. It is partially shaded and very little sound filters in from the surrounding bush. Half-way up the trunk of one gum tree, a rusted horseshoe partially protrudes in mute evidence of 'what' or 'who'?

Once part of a vast area taken up by The Barrow Creek Pastoral Company, an outstation homestead was built there, the outlines of its ruined walls almost completely mingled with surrounding stones.

Tragedy struck one night when native spearmen surrounded the house and ignited the thatch roof with fire-sticks. There were two stockmen inside. One of them rushed out and was clubbed down while a third man hid beneath a bed until all was quiet and he was able to make his escape. There were cattle in the stockyard and all hands were engaged in branding when the survivor struggled into the main homestead. In their hurry to get to Anna's Reservoir, no one thought to let the cattle out of the yard and for many years a deep trough scarred the inner perimeter of the stockyard where thirsty cattle had circled.

Blackfellow's Bones, a site on what afterwards became Mount Riddock Station, marks the site of the reprisal which followed. About ten men women and children were indiscriminately shot by a police party under the leadership of Mounted Constable Willshire.

While at Anna's Reservoir with their sheep, Billy, Clara and the children would camp on the opposite side of the creek, distrusting any anger which might still lurk spiritually among the fallen masonry. The waterholes they had and which they used alternatively, were always places of special significance in Aboriginal lore and were treated with the utmost respect. Always in rocky places, tales were told giving personalities, names and special significance to every landmark. As his reading extended, Ben developed an abiding interest in warfare and battles fought down through the ages. In his spare time he would fashion armies of clay warriors and position them. As steeped as he had become in Aboriginal mythology, equally the ancient histories of distant places glued him to his reading.

Building the Homestead

There was a good supply of local clay and it was used to bind the homestead walls, together with stone quarried from the nearby Hann Range. When time permitted, Sam dug limestone at a gap nearby, carted it back and burned and slaked it in a pit beside the well.

It had become obvious that their temporary camp was going to be permanent and every spare moment of their winter months was put to good use preparing to build. Liz would take any of the children not otherwise occupied to the hills with a picnic lunch, where they would dig and loosen suitable stones and pile them ready for removal to the building site.

The South Australian Government requested two favours of them due to their position on the Overland Telegraph Line. One was that they be prepared to water passing stock including large mobs of cattle moving overland. The pay was a penny per head. The other was to serve as a Post Office agency. Once more solidly housed and weather-proofed the Overland Telegraph would install a telephone linking them to the rest of the country by relay along the line.

Travelling south from Darwin, Mr and Mrs Norman Octoman of the Overseas Cable Company were induced to stay awhile. Norman was suffering from tuberculosis. They stayed six months and the dry Centralian air visibly improved the patient's health. Because he had formerly been a stone mason, under his direction and inspiration, Sam and the family laboured willingly and the homestead finally became a reality. As a parting gift, Norman took both ends of a kerosene-tin box, sanded them and created a pair of pictures in the sepia/browns of the available house-paint. He was a talented man and kindly and the same must be said of his wife, who used her sojourn to teach the younger children and filled in some of the cracks in the 'book learning' of the older two whenever time permitted.

The Sandover Alligator

Bob Purvis, a powerfully built young man from Queensland, arrived in the district travelling as camel-man with Mr Plowman, the 'Man from Oodnadatta,' the overlanding parson.

With camels and wagon, Bob then turned his hand to the job of carting iron poles to replace the ant-ridden wooden ones. By the time that task was completed Bob had fallen under the spell of the Inland and was bent on saving enough money through well-sinking and carrying to eventually take up his own piece of country. He found it to the east of Glen Maggie and named it Woodgreen.

The very names pioneer settlers chose for their pastoral properties appeared an obvious indication of their longing for the fresh green growth of their homelands and an underlining of their determination.

Bob spent a week putting his wagon, together with his mighty strength, to good use. He concentrated on carting the stone already organised by Liz and her team and when his turn came to build a home, they in turn went to help.

He boasted a prodigious appetite and because he had come into the Territory initially by way of the Sandover River had become affectionately known as 'The Sandover Alligator.' He managed to impress the children with tales of his prowess. He could even 'put up a hundred fence-posts a day and walk a mile between each one.'

It was obvious that in the very scarce population surrounding the children, everyone was a hero.

Bush Kids

With the adults and his older brothers so busy, Ben was often left in the role of child-minding his sister Mags, two years younger than he was. He wanted so much to wander off and look at everything fringing their living space, but Mags wasn't able to keep up with him. He studied the problem and found a solution which suited him far more than it did Mags.

He would carry a lengthy piece of cord and a spare billy-can of water. Once out of sight of parental control, he tied Mags to a shady tree and leaving her with the spare water, would take off on his wanderings.

Mags didn't seem to object too strongly and he managed to get away with it for weeks before someone found the tracks which in themselves told the tale. Mags suffered a few minor inconveniences like ant bites but she learned to amuse herself while she patiently waited for Ben to return and take her home. She never thought to report their activities because Ben was persuasive and encouraged her to believe it was all part of a game. Perhaps he was a 'Red Indian' and she was his captive, or he was more frequently a soldier taking prisoners of war.

Because twenty pupils were required, when Mrs Stanley arrived in Alice Springs to teach, Mags, then three, had been roped in to help make up the number. Reading came easily to her and being left under a shady tree with a book was for her no major hardship.

There was a tree she climbed once, and an abandoned birds nest with three pretty eggs nestled in it. Mags retrieved them and stuffed them down the front of her dress but fell part-way down and crushed the eggs. The smell was decidedly revolting because the eggs were well past use-by date, and when Ben took her home they were both banned from entering the house. It took a lot of soap and water under a hose thrown over the lip of the water-tank to mend the situation and Ben learned to keep a more careful eye on his little sister.

An essential front verandah was added to the homestead, affording afternoon shade. It became a multi-purpose area sometimes given over to travellers for sleeping quarters, but most times a place where Liz liked to sit and sew and mend. She bought rolls of fabric from passing Afghan hawkers and striped blue bombazine happened along more often than did any other material. Her children would later remember that striped blue bombazine in many shapes and sizes flapped along many outback clotheslines in that particular decade.

From her verandah she could see the well and watch stock coming in to water. She could view a wide arc of grey-green mulga trees and beyond them the hills cradling their little valley. Scuds of dust foretold the return of one of her family from track-riding, or fencing or perhaps from water- or wood-collecting. Swoops of budgerigars and zebra finches would chitter in low waves across her vision and crows gossiped and complained. Slow-moving hawks circled high in the distance pretending none of this activity was of any interest. Sometimes she watched a dingo slinking in to water at the trough, looking over its shoulder nervously, trotting from shade to shade in the vain pretence that shade made it invisible.

From her vantage point she was able to keep an eye out for any raiding goanna seeking her hens and their eggs or chickens. Her fowls were precious to her and she was at constant war with marauding wildlife.

Crows were a constant menace and when Ben found one which had trapped itself in the fowl house, he wasn't going to simply let it go. He tied a long piece of string around its leg and thrust it under a box until he could put his plan into practice. Under cover of night, he dipped a piece of saddle-cloth in kerosene and attached it to the other end of the string. He lit it before letting the crow fly free and ran into the house shouting, 'Look at the comet! Look at the comet!'

The family hurried out to see the comet but what met their eyes was a series of spot fires. Bothered by the string, the crow flew awkwardly and kept landing, every touchdown starting another fire.

The family rushed for bags and fought the flames but as soon as one fire was beaten, the crow landed and another one began.

Sam's voice could be heard issuing out of the smoke-filled night, 'Look at the comet, look at the comet, I'll give you 'look at the comet,' Benjamin Esmond Nicker!'

Ben went bush for two days while Clara sneaked a blanket and food to his hiding place.

The crow could be said to be easily the most unpopular member of any bush community. He was graceless, raucous, he picked the eyes out of new-born lambs, and stole the eggs out of the hen-house and there were so many of him. He seemed never to be anything other than full-grown and baleful.

When Gene and Ben trapped a second one, they thought carefully before they put any plan into action. This time they tried a trick they had heard about. They tied a piece of bacon fat to a very long string and fed it to their captive before turning him loose. He picked at it before he swallowed it and it wasn't long before the fat passed through still attached to the string.

A second crow investigated with the same result. They had seven crows threaded between the pork and the post to which the string was anchored before they were discovered and were happily enjoying the clumsy, frantic ballet of their prey trying to free themselves.

Sam didn't quite know how to punish them for this offence. He hated crows. He gave them two days to think about what they had done, by which time the two miscreants had talked themselves into a state of guilt and their father had seemingly forgotten all about it.

To understand the bush, one had to have first absorbed the basic knowledge of all its creatures. To track an animal was more simply done if one knew its habits, its habitat, its appetites and generally the blueprint of its everyday living. This was an area which had been taken care of by Billy and Clara while the Nicker children were too young to realise they were being taught. It was also an absorbing method of whiling away child-minding hours in the only way they knew. Together they gathered everything edible, bush fruits, nuts and berries, yams and yulkas, avoiding those things which were poisonous. They learned 'finger talk' and to watch the behaviour of wildlife and to know when rain was coming or dry times threatened.

At the same time Sam explained cloud formations and scientific reasons for the subtle variations in the colour of the sky. Perhaps their earliest lessons taught them that there were two answers to everything and when they thought about it, each solution in its way helped explain the other. Growing up, both backgrounds made perfect sense to them and they were able to switch effortlessly from one to the other. When they spoke in any

Aboriginal dialect, they thought Aboriginal. When they spoke their mother-tongue their thinking became automatically European.

'That one wee-i all about and that one quee-i, they talk alla same Blackfella.' Billy liked to announce. But most outback children adopted the Aboriginal dialect of their own particular area and those who did not later regretted their lack. For Elizabeth and Sam, and for Billy and Clara, each adult couple was too firmly ingrained to learn more from each other than was required, and that they adopted no more than a pidgin English when conversing reveals perhaps their subconscious barriers. But there is no doubt that they liked and trusted each other.

Sam the Handyman

Despite the everyday business of establishing a pastoral property and raising a family, a kitchen, dining room, bathroom and extra bedrooms mushroomed behind the main building and the Post Office/store. They were erected by a professional builder, Tom Williams, and cut from local stone. The roof was corrugated iron overlanded from Oodnadatta and while they waited two years for the roof capping, rubberoid served the purpose. Their floors were flag-stoned. One thing never planned was a lockable door. It wasn't ever considered in those days because one never knew when a traveller might happen along needing a place to sleep or a meal or even medical attention. If there happened to be nobody about to welcome them, there had to be free access to what comforts were available. It was outback hospitality.

Sam rather fancied himself as an expert in the making of beer. When a traveller was really desperate, Sam would be happy to sell him some of his home brew and there were keen return-customers until one particular batch almost ruined his reputation. Liz had been peeling onions in a bowl and because water was so carefully hoarded, she strained the onions and left the water in a large jug. When Sam took over the kitchen for another batch of home brew, the end result didn't go unnoticed among his customers.

'Next time Sam, go easy on the onions!'

Grog wasn't a commodity easily come by. It was a known fact that in desperation, some of the more isolated old 'codgers' licked strychnine corks for a kick. There were a number of these men, some of them failed prospectors and some remittance men. A few escapees from maintenance payments, and some who had simply turned against society and moved camp around the outback. Some found jobs looking after outstation bores. They settled down and grew watermelons, pumpkins, tomatoes and whatever vegetable seeds they could get hold of and were content to be left alone except for the monthly supply visit from 'the boss'. They usually acquired a dog or two for company and one old man, grumpier than most, was roundly criticised for hobbling his dog's front legs. Even his dog would have left him if he could.

In those years everything came by camel-train and it took many months for any ordered item to arrive. Footwear for growing children was a major problem as the children outgrew their shoe size while waiting, so Sam tried his hand at boot making. Beginning with a sort of Roman sandal, he progressed quickly to boots and despite the children's obvious reluctance to wear them, Liz insisted. This was an area where she was adamant and it might have stemmed from the constant mending of damaged feet as much as from her sense of dignity.

Bartering

There was little hope of any kind of a garden. The well water had a high mineral content and was therefore too 'hard' to grow anything except a few pumpkins and weedy tomatoes which struggled along the bathroom overflow. Instead they managed to plant and protect with fencing, a good supply of staple vegetables at Conner's Well. A round trip of a day's ride, it was customary to do a little gardening while checking stock and water there and a simple matter to reap whatever was ready to eat.

Potatoes, onions, rice, flour, tea and sugar constituted their mainstays, as well as kerosene which, when carried in four gallon tins by camel, leaked into the foodstuffs. It was many years before the children realised that everything didn't normally taste of kerosene.

Axle grease was home-made from linseed oil, castor oil, tallow, resin and beeswax. Salt was collected from a salt-pan at Coniston, their western neighbour. Soap was home-made in their copper. Their little store did well and accounts were sometimes settled by barter. An entry in the Glen Maggie store journal reads... 'May 18 1926. S.F. Nicker settled all accounts with T. Moar. Moar now owes me 91 head of mixed cattle and 10 new wool bales.' Cattle prices were dismally low and further down the same page is the following entry... 'Tom Moar Account. 58 head of cattle at 2 pounds 10 shillings, 145 pounds.

Account, 112 pounds, fifteen and sixpence. 10 wool bales, 5 pounds.

Result equals 27 pounds five shillings and sixpence.'

The best price the Nicker's ever got for wool was not worth the effort involved. In the aftermath of World War 1, beef prices rose dramatically due specifically to the fact that Australia won a contract to supply beef to the Belgian Army. Glen Maggie and pastoralists Australia wide concentrated on cattle, but by 1921 when the contract ran out, prices dropped so alarmingly that on several occasions Sam had to pay to cover agents fees, transport costs and sales commission. Prices had dropped as low as two pounds ten a head for bullocks.

Their sheep proved to have been worthwhile after all because now when finances were at a low ebb, people who were tired of eating beef were happy to indulge in mutton for a change. Without any financial backing from banks or stock agents and despite their paucity of water the family were, despite all odds, not only surviving but making a pretty good living.

Stockwork

Perhaps inspired by what they thought of as the glamour of riding horses, a number of young Aboriginal lads wanted to learn stockwork, but Sam's horses were precious and he refused to risk them. In his opinion, coming straight out of their bush camps, the local Aborigines were not able to sit a chair and therefore unqualified to sit a horse. To him, they had no understanding of animal care. He had already witnessed too many saddle sores and ruined mouths among stock-horses on neighbouring properties. In later years, Aboriginal stockmen earned themselves a place in outback history but horses were too precious and stockwork was still within family capabilities in Sam's lifetime.

As soon as they were old enough, his own children rose early and followed him outdoors with bridles. They stood in the freshness of a new, dimly-outlined day and listen carefully for the sound of distant muffled horse-bells. The working plant was always turned out hobbled overnight to graze and it was unusual for them to forage far away. Walking through the waking day it was a secret world and a time for the quiet conversation children love. Sometimes they paused, listening and altering their direction towards sounds of jiggling hobble-chains coupled with the clinking of the neck-bell.

The browsing horses barely lifted their heads while Sam and the children each approached a different mount conversationally.

'Come on old girl, you've had a good feed. Time to get back to work. Steady on now. Let's put your bridle on. Gently. It's not a bad day and we've got plenty to do. Won't work you too hard. That's a good girl. Let's get those hobbles off and get going.'

Once mounted, Sam watched and advised his children while they scrambled on board as best they could. It wasn't long before they were able to bring in the horses without adult supervision and it was a job they loved. Probably one of their first independent accomplishments. 'Doing is learning' and 'Watching is learning' were two of Sam's many maxims and the children learned to observe and to follow his lead with horses.

For fear they might have a foot caught in the stirrup and be subsequently dragged, the children initially rode bareback. Once they had comfortably developed a 'good seat' and 'hands', they were elevated to either a riding-pad or an army saddle. They were natural horsemen before they received the privilege of a 'proper' stockman's saddle. It was a cardinal sin to give a horse a sore mouth or a saddle-sore and the wearing of spurs was strictly prohibited. 'If you can't get the best out of a horse without spurs, you can't ride.'

Track riding was always one of the station priorities. There were no fenced properties. Fencing wire was difficult to transport in the vast amounts required and the expense would have been prohibitive. Instead, it was an almost permanent job riding tracks, especially when rain had fallen and stock followed storm water. In this part of the country cattle always walk or browse in a south easterly direction into the prevailing breeze. It was a commonly accepted fact that every pastoral property fed on the cattle wandering through from the north west. Nobody ate their own cattle and it was thought almost indecent to break the chain. When track riders or musterers crossed your borders, it was polite to inform the property so that attendant riders could join the stock-camp and keep tabs. Stock-camp tracks crossing your boundaries without authorisation had to be well explained if goodwill was to be maintained. And goodwill was an important commodity. On another man's station property it was not considered acceptable to attend the killing of an eating beast. If the hides were kept for making hide ropes, the brand and earmarks were always cut out and disposed of. There were many jokes circulating about new chums asking the silly question, 'Why do you cut the ears off?' The stock answer was usually, 'Why, to make them bleed better.' When a pastoralist was asked what his beef was like, it was acceptable to answer, 'I wouldn't know. I've never eaten it. But my neighbour's is very tasty.'

It was an age when straight backs were demanded of children and learned while they sat at table. To slump at mealtime could invite a broom-handle slid through the collar-line down the inside of the shirt, out through the waist-band to the floor. Parents induced the child to endure while everyone else sat erect and avoided eye-contact. Slumping, slouching and bad manners were punishable offences but all of this posture training was excellent for horse-riding. Slouching in the saddle was guaranteed to cause saddle sores for the horse and put him out of commission. It wasn't good for the saddle either and if the horseman came off with backache then he deserved it.

Mustering, branding, shearing, track-riding and the perpetual movement of stock to and from waters kept everyone busy. It was often a case of all hands on deck and particularly at branding time. Mags never forgot a day

Sam Irvine, Royal Mailman (above).
Sam driving Royal Mail leads a convoy, 1927 (below).

A costean dug during the 1933 Terry expedition.
Photograph by Bill and Bernie Manley.

Yintardamurru (inter-amoru) rockhole near Mount Singleton. Terry's journal states, 'Tuesday August 9th 1932 ... I returned to find Nicker had found a large rockhole with a sand soakage in the bottom sufficient for our needs ...'
Photograph by Adrian Wynward Smith.

Paddy Jupurrula and Darby Jampijnpa pointing out Nicker Rockhole with Adrian Wynward Smith.

16th March 1927. Ben's signature beside a soakage near the Granites.

'Eaters of Stone'. Stan O'Grady and Ben Nicker dollying samples on a cold winter morning. Western Desert.

Bob Buck (extreme right).

The Mark Foy expedition, 1936. About to leave Alice Springs.
Photograph H.V. (Bill) Foy.

when she was half grown as was the heifer she was allowed to rope by herself. There was a post in the yard and in theory she had to quickly loop her end of the rope around the post so the animal could more easily be pulled up to the slip-panel, where further ropes secured it for branding. But Mags and the heifer ran in an anticlockwise direction around the post, each trying to escape the other but in fact with every appearance of chasing one another. Billy's voice soared above all other sounds, urging her to 'Loog out, Magginick!' Maggie Nicker was 'looking out,' as fast as her legs could carry her but in her panic it was awhile before she remembered to drop her end of the rope and break the circle.

The homestead stockyard wasn't a great distance from the house and neither was the 'dunny.' The 'dunny' didn't boast a door, instead a bough-shed wall was built a discreet few feet in front of the opening. A handy stick, 'Mum's shillalagh,' leant against a post. This was the anti-snake weapon. A looped wire held bunches of 'toilet paper' torn or cut from outdated newspapers and Stock and Station catalogues. Enthroned there one day while the men and neighbours were 'cutting out' a few cattle at the yard, Liz was startled when a bullock sped between her seat and the sheltering wall. It was immediately followed by a stockman she had barely met. 'He was very gentlemanly,' she announced afterwards. 'He raised his hat.'

Sam's Cats

Bush mice were always a problem when man bought foodstuff into an area and knowing this Sam had introduced a couple of kittens to the household. He was fond of cats and the naturally extending feline family followed at his heels when he was home, well aware of his affection and ever ready to spring to attention, should he pick up the meat knife and sharpen it against the steel. They could be certain of a trip to the meat-house and of the special tit-bits tossed in their direction. They followed him too, when at night the family were bedded down and Sam would habitually do his last patrol down the slope to the well to check the trough level and ensure the float was securely locked. To this very day, many Aboriginals remain reluctant to camp there because they are convinced that his ghost remains.

'That one oldman, he bin walk longa that one well, eberynight he bin walk that way with that one 'urricane lamp and that one big mob pussy cat, walk longa im.'

Liz Takes a Tumble

The well was first priority among the day-to-day jobs at the homestead. When the boys were absent on any other part of the property it was Liz who attended to its care and upkeep and who would have to handle the watering of passing mobs of cattle. It was a two-man job. One would lead the horse or camel up and down the race raising the buckets, while somebody else would have to lift them across and tip them into the chute and thus into the tank. When the menfolk were away, they made certain the tank was full because Liz would never have been able to manage it.

The well-head was on top of a high, stone structure like a pyramid cut off half-way. The climb was steep and difficult for a little lady wearing cumbersome, hitched-up skirts but she managed it whenever she had to, until came a day when with a high wind bothering her, she missed her step and fell.

There was nobody else home at the time, and she, who was of such value to everyone else when they were hurt, wasn't now easily able to help herself. It was obvious she had broken her ankle and the house was quite a distance from the well. The pain made her thirsty but she couldn't reach the water. There was nothing within reach she could use as a crutch and apart from tearing out the sleeves of her 'shirt-waister' and bandaging the ankle she wasn't able to think of any way out of her predicament. She hoped one of the quieter horses might come in to drink then she could perhaps coax it to her and clamber on, but the few who watered at the trough eyed her askance and pranced away. She wanted to sleep but pain kept her alert as did the difficulty of her situation.

As the sun lowered, she found herself shaded by the structure of the well and that brought a measure of relief but with an added anxiety because she feared the coming night's chill would intensify her pain.

She must have dozed. A smell of wood-smoke seeped into her consciousness and she pulled herself to a sitting position. A pale glimmer of lamp-light shone through the window of the house but it was too distant for anyone to hear her if she called. As she watched, the light seemed brighter and a hurricane lamp moved through the front door and bobbed in her direction while Sam's disembodied voice called her name.

Relief!

There was no way to move her up to the house other than by wheelbarrow and this was an instance when pain overcame her dignity. The wheelbarrow was a cumbersome home-made one used sometimes for the ferrying of wood, or removal of manure from the fowl-house. The barrow's many mainly aromatic purposes gave the occasion a sort of embarrassed drunken air but Liz, overdosed on whisky, Sam's universal panacea, rode as gratefully regal as her circumstances allowed.

For a few days, Sam took over the housework. He set the ankle and carved Liz a crutch and fussed. He complained that every recipe stated 'First take a clean bowl....' when he'd already run out of clean bowls. His workman's hands couldn't break an egg without a lot of swearing. He forgot to feed the fowls who consequently complained stridently around the house and flapped in through the door when it was imprudently left open.

The result was a great deal of running about with a broom trying to remove them, unsuccessfully, while others found their way in through the open window. Flies swarmed into the kitchen. When Liz opened her eyes on the third morning and found a goat eyeing her while it happily chomped on one of her petticoats, she decided enough was enough. A housewife Sam wasn't.

Her holiday was over.

Verandah Viewpoint

The country around them grew better with every wet. From the homestead vantage point Liz noted an improvement in grasses and a slow but steady greening and developing density of shrubbery. Because she was at heart a gardener, she believed the stock they had were responsible. Their hooves broke up the top surface and their bodily waste nourished the soil. Where they foraged on low bush branches, the canopy grew taller and shaded more grass and infant trees. Moving away from their watering places, they distributed grass and herbage which better anchored what already grew. Every hoof indent left a cradle for new seed to develop, protected from winds on the open plains and held little pockets of water when it rained.

Her own little stony ridge was unproductive and the seasons undependable but the propagation happening around her became a fascination.

She picked no wild flowers for decorative purposes, for which she might have been excused, but left them where she could watch them grow, deriving more pleasure accordingly. She also felt that picked flowers must be faithfully attended to, and she had little enough time to spare, else they might wither in their vase and consequently someone she knew might die. Indeed the superstitions, brought with her parents from overseas, found fertile soil in Liz's isolation in the centre of Australia.

She was finally able to unpack her most precious belongings brought all the way with them on board the buggy from Queensland. Her exquisite china dinner-set graced heavy home-made shelves along one wall. A whale's eardrum sat above the open fireplace in their front room exciting interest from the children and visitors alike. A finely crafted and highly decorative rug from Samarkand graced one wall, its colours slightly dimmed with travel and too precious to tread underfoot.

But their most valued possessions were their books. They were kept stacked upon a table in one corner and teetered against encompassing walls interspersed with magazines and too well-used to ever be neat. To protect the most avidly-read, Liz sewed saddle-cloth covers when that fabric was available and sometimes patchwork or blue bombazine, whatever she could spare. Her efforts served her purpose but the result was inelegant and she refused to apologise for the perpetual disorder in that corner.

When it was suggested they were moving further away from civilization, their reply was that they took civilization with them, never realising that times elsewhere had changed.

But the children had become accustomed to their lifestyle. Wherever they had lived their lives were enriched by occasional travellers who called and stayed awhile. They spoke of other places and other countries, of wars and famines, snow and ice, of air planes, cities and multi-storey buildings, picture-shows, ice-cream, the sea and sailing ships and all the marvels bush children could only read about. They brought the outside world to life. The time had come to take the children south.

Outback Kids Visit the City

By 1916 the road was better travelled and no longer carried the risks it had. Billy and Clara were established and good rain had fallen. The cool winter months were an excellent time to travel and there was business requiring attention and relatives to visit.

Nearing the Adelaide skyline, the children's excitement knew no bounds.

A breeze tendered an aroma somewhat unpleasant.

'The sea! The sea!' they cried in unison. 'No,' they were informed. 'It's the sewerage works!'

When they finally saw the sea at Glenelg, they were speechless. It defied the imagination. They visited the Museum, Art Gallery, city parks and the Zoo; strolled through city streets and wandered in the hills overlooking Adelaide. Electricity and ice. Balconies and stairs. Paved streets and shop windows. Motor cars and trams. Cold grey rain and overcoats. A lot of people who didn't want to stop and pass the time of day. 'Adelaide strangled my mind,' said eight-year-old Ben.

They went by train to Melbourne where the weather was wetter and colder and the seaside even further away but they were met by relatives of Sam's and passed from one to another, among them cousins nearly their own ages but with whom they found nothing in common. Despite the best efforts of both parents it was obvious they were raising bush kids. Thinking a solution might present itself between the relatives, Sam and Liz resolved to leave the older boys to get a few years' more formal education than they had already received but the boys refused in no uncertain terms.

Perhaps in a few years' time Ben and Mags might be prevailed upon.

Ben had ideas of his own and schooling in Melbourne was definitely not on his agenda. Having found and tasted the feverishness of cities, the brothers rejected them as 'nice to visit but wouldn't want to live there'. Mags, however, was younger and delighted with all she had seen and would return when she was judged old enough, to Melbourne, her relatives and the furtherance of her education.

Droving

Meanwhile Ben prowled the perimeters of Glen Maggie with his camel and came to know everything there was to know about that part of the world. Sam having twice extended the property boundaries, there was now a larger backyard to find out about. Ben acquired two more camels and taught them to work the whip at the well, pulling water. He trained them to harness with the wagon and used them to haul water from the rockhole twelve miles away, still the family's only drinking supply. Claude and Gene left him to his own devices for they were more content with their horses and were developing reputations as jockeys at every race meeting within their reach.

When stock prices warranted, the boys were happy to undertake the long droving trips to Oodnadatta. It was a welcome break in station routine. They were sometimes augmented by friends or relatives from other properties with additional cattle and other horse-plants. These trips were an opportunity to catch up with folk along the route as well as companionable working time spent with neighbours. Distances between properties precluded much opportunity for young people to get together, therefore when given the opportunity, market cattle were mustered and headed out with great enthusiasm. Droving could be tedious but young people found methods of breaking the monotony when everything was going well.

There was always the risk of the unexpected. A mob of seemingly contented cattle could rush at the shadow of a sound in the middle of the night and plunge heedlessly through scrub too low-growing for a horseman to easily follow. The night-watch had to, any way he could, race to the lead and shoulder the lead beast back on its tracks so that the following mob would follow and circle until it settled. Ringing. Night horses were especially valuable animals and the partnership between horse and rider under these harrowing night-time circumstances was extraordinary. It was as much the horse's senses that saved a rider from losing his saddle under a branch, or that helped him to keep his seat when his horse narrowly avoided a dangerous misstep in dust-black night.

Every day presented a different set of challenges until routine settled down the herd. Weary riders overlooked perpetual dust, prickles in swags and sweat blister and the body adjusted to the lack of baths. Everyone smelled like horse-sweat.

Damper, with corned beef served curried, stewed or simply in cold slabs with pickles and a pannikin of black, sweet tea. Anything tasted good, especially 'Johnny-cakes' cooked on the coals and slathered with golden syrup.

A few 'killers' were included in any travelling mob providing fresh beef along the way and although much of it had to be salted to preserve it, for a few days fresh steak provided a gratification dependent on the cook. A camp-cook who could ruin steak might well have found himself relegated to horse-tailer, or sent packing with the very next passer-by.

At Oodnadatta, journey's end, the mob would be yarded and trucked to Peterborough for sale. One of the men went with the cattle as train drover to attend to the watering and feeding en-route while the rest of the team enjoyed the township and companionship for a few days before returning with their horses to home base. It was 300 to 400 miles home, depending on the particular property they started from.

'Trollop's Pinched the Jam!'

Eugene, or Gene as he was better known, was a gentle soul. He never spoke until he was six years old. He suffered, it was thought, from a form of St Vitus Dance although without proper diagnostic assessment it is doubtful. Whatever the problem, he outgrew it at an early age but was ever afterwards considered not as strong as he should have been. As is usual with children, however, he never let ill-health stand in the way of doing anything he wanted to do. His schooling wasn't as intense as his siblings, not withstanding he absorbed their knowledge despite the fact that he was never a penman. He was probably more at ease in the saddle than on a kitchen chair and less able to socialise than they were, but enjoyed company and was that rare thing, a good listener. He was an affable man but perhaps a little lonely, a little withdrawn due to both his early illness and the shadow cast by his brothers. He would never have believed his quiet pleasantness made for a relaxed companionship.

His first words were recorded in family memory. Mags wandered into the kitchen once, covered in dirt and mud and Sam in exasperation remarked, 'You look like a real little trollop'.

Some weeks later Liz bottled a large batch of jam and left it to cool. Two-year-old Mags investigated and found it so tasty that she had demolished quite a lot of one of the jars. When the family gathered for dinner they were informed by Gene, 'Trollop's pinched the jam.'

While everyone paused in amazement, he insisted, 'and it isn't the first time either!' For the next two weeks the poor lad had to bear the nickname 'Willy Wagtail', for everyone knows that bird's habit of reporting everything he sees.

Mags was never sure whether she favoured horses or camels. In her estimation each had its advantages depending on the job at hand. Sent to Conner's Well to check the water supply or the animals there, she would choose to go by camel, able to catch up with her reading while she travelled. A camel saddle is a double-framed affair to better sit the hump and the front part provides an excellent book-rest. She well understood Ben's preference.

A camel-man more often walks and leads his camel, mounting only now and then to rest his legs. A camel's gait is not conducive to a comfortable ride for any great length of time and camel knee-joints suffer from the effects of too-frequent hooshing down. It isn't any wonder they grunt and complain so much.

When she had different tasks to undertake, like mustering, branding or track-riding, Mags was as equally at home on horseback. When the family attended local race meetings she rode in the lady's race and sometimes won. There wasn't a lot of competition. Sam encouraged her and Liz kept her thoughts to herself.

3
'Little Bit Long Way Benninick'

The Coniston Murders in the Tanami Desert: 1928

The Aborigines of the Tanami Desert were unfriendly, as evidenced by their sweep across the country five years later, a story well documented as the Coniston Murders. For many years after the turn of the century, trappers, miners and prospectors had come under attack, often with fatal consequences. In 1928 Fred Brooks, a seventy-year-old dingo scalper, camped at a soak on the Lander Creek, was clubbed to death and his dismembered body stuffed down a rabbit burrow.

Alec Wilson and two Aborigines were travelling towards Coniston with Joe Brown who was ill with beri-beri. In urgent need of medical help, time ran out for Joe and his friends buried him in a shallow grave until they could get to Coniston and return with a wagon for his remains.

On their way they discovered parts of Brooks' body. Arriving at their destination they sent a message alerting the police officer stationed at Barrow Creek. Returning to Joe Brown's grave, they found the body disinterred, stripped of clothing and all his possessions stolen. The police patrol investigated Brooks' murder site and duly arrived in Alice Springs with two men charged with Brooks' murder. While they were busy, Nugget Moreton, a cattleman who was camped about sixty miles from Coniston on the upper Lander River, was attacked. He fought valiantly and survived but at the cost of a pulped chest, broken ribs and shoulders and seventeen pieces of boomerang imbedded and later taken from his head. Despite such horrific injuries he managed to catch and mount his horse and ride to Coniston for help.

The police party returned. 'Tipinpa' is a soak between Mud Hut (where Nugget was attacked), and Coniston Station and it was here and in the surrounding area that a cohort of settlers under police leadership took their revenge.

It was an ugly episode in outback history, the full facts of which may never be known or truly understood. A traveller, coming upon the site of the major massacre, afterwards reported that as many as sixty skulls, lying like bleached melons, littered the plain.

An official enquiry into the killings led to the dismissal of the police officer concerned and merely a sharp reprimand to those who rode with him. They were marked men among ordinary settlers and never again regarded with a great deal of respect.

Ben Crosses the Tanami Desert Solo: 1923–24

In 1923 when Joe Brown, explorer and prospector, called on his way to cross the Tanami Desert and reach the north-west coast, Ben prevailed upon his parents to let him go. Because Joe's horses wanted nothing to do with his camels he was forced to leave them behind and take instead two of the station plant, one to ride and one to pack. Joe proved to be almost impossible to get along with. Described as an argumentative and difficult man, perhaps he had spent too much time prospecting in isolation. At fifteen, Ben was still of an age to respect his elders, but on reaching Hall's Creek man and boy parted company and Ben, with his own two horses headed back towards Glen Maggie alone.

The Coniston murders were yet to happen but in 1923–4 the inland was in the grip of drought. Because bush tucker was affected, the Walpiri Aborigines were spearing greater numbers of stock and ill-will was escalating. Ben took every precaution. Where he camped at night he lit the most meagre of fires. He would hobble out his two horses, place packs and saddle alongside the campfire, then take his swag to a distance so that any night attacks would be deflected. He knew how to follow birds to water and where to find bush tucker. Although in scant supply there was always a feed to be had for a lone traveller who knew the bush.

Moving across a spinifex desert can be difficult. It isn't possible to walk or even ride in a clear-cut direction because one must weave a path between needle-point clumps which can spread to almost twenty metres across. Where they pierce the skin, blood poisoning can set in quickly.

Leather leggings are a help but there are ringed spinifex hummocks sometimes hip-high and hostile. Unless one fixes the eye on a definite point ahead, and that's not a simple matter in the sameness of a desert, the constant weaving can easily disorientate a man.

In the 1890s, gold rushes to Tanami and Hall's Creek were hazardous and many men perished. Gold always seemed to be discovered in inaccessible places. Subsequently the Western Australian Government provided a string of wells across the desert. By 1923 when Ben first crossed the Tanami Desert, the wells, like the diggings had long fallen into disuse. There was little sign of habitation at Tanami but a relative township still standing at Hall's Creek. It was a collection of mud-brick buildings, post office, pub and store and a handful of homes.

The Great Sandy and the Tanami Deserts merge south of the town and present a daunting reality. The sand plains of hummock grassland (spinifex) stretch for hundreds of miles in flat landscapes peppered with stunted mulga groves and acacias. No surface water exists except immediately after rain. Citadels of termite mounds rise up here and there and tiny kangaroo mice, called alupi, scuttle from view. The Rufous Hare wallaby blends with the colour of surrounding sand and scorpions, ever offensive, assault the unwary, but the desert is above all, the natural home of the lizard. It comes in all shapes and patterned colours from deceptively lounging monitors to the minuscule and Ben could survive, when he had to, on lizards, munyaru plant and yulkas, tiny onion-like root vegetables with a potato taste. All Billy and Clara had taught him had been well ingrained.

Anne-Jane arrived at Glen Maggie by car looking for Ben.

She had dreamed of Ben coming out of the scrub from west of the well. Worried that something serious might have happened to him she drove north from Undoolya Station to share her anxiety. While she spoke with her mother and step-father, Ben did indeed ride in.

At fifteen and without a compass, he had crossed the desert alone. Where many men had perished he had triumphed. He had covered perhaps a zig-zagging distance of seven hundred miles from Hall's Creek to Ryan's Well.

Michael Terry, in his book, *The Last Explorer*, wrote of Ben's remarkable trek:

> 'At fifteen, he had come back alive and well. He had safely completed the finest, riskiest, solo venture in Inland history, so I claim. I found Ben to be the steadiest chap, deeply knowledgeable in bush lore of every sort. To him, to be a bushman was not just a question of instinct so much as observation and remembering. No two trees, no two ant-beds, no two hills were to him exactly the same; all was recorded in a mental picture.'

Driving Men Mad

Anne-Jane was arguably the first woman driver in the Centre. The car was a 1922 Dodge utility and one of the first two resident motor vehicles. It was a measure, not only of her expertise but also of his confidence in her that Ted, her husband, had entrusted the car to her. For himself he preferred horses and would never enjoy driving the contraption. When he did, he would talk to it, always amazed that it would never, like a horse, respond to voice command.

On his first attempt, with the physical exertion of his bellowed 'Whoa, there!', his foot clamped down on the accelerator and the car bolted enthusiastically through the scrub while he glared helplessly at the controls.

His passengers turned quite pale and having with difficulty released their grip from wherever they clung, tumbled out and elected to walk.

Ted's neighbour, Louis Bloomfield of Love's Creek Station and perhaps more mechanically inclined, invested in a motor and proudly drove it into Alice Springs. He did very well until coming over a jump-up a little too quickly, he lost control and ploughed into a tree. He was certain the tree hadn't been there before and subsequently the two friends agreed the motor car would 'never catch on. Too dangerous'.

Louis Bloomfield

Louis Bloomfield was a horseman of note. He worked his way across southern parts of the continent buying, selling and breaking-in and did stockwork on cattle stations in return for horses. While he was working in a stockyard on a property near Oodnadatta, a bullock with sweeping horns spun and gouged him in the stomach. As it plunged, the horn tore the initial opening, gashing it so widely that Louis' bowel was dragged out. An Aboriginal stockman working with him carried Louis from the yard bundling his entrails as carefully as he could. While someone saddled a horse and raced to Oodnadatta, seventy odd miles distant, Louis was wrapped in rugs and given whisky. Mrs Kunoth locally known to everyone as Granny Kunoth, was to the southern district the counterpart of Liz Nicker further north. Both untrained but imbued with a lot of common sense they each served as midwife and nurse whenever and wherever required.

Granny Kunoth hitched up her buggy and grabbing everything she might need, hurried to the scene of the accident. Because of the distance, it was a couple of days before she got there but the patient was still alive. She washed out the cavity and the bowel with a solution of Condy's Crystals and carefully removed every piece of stockyard debris. That done, she pulled hairs from a mare's tail and boiled them together with scissors and darning needles. Her next task was to push the bowel back into the wound in the correct order. The patient was heavily sedated with copious amounts of whiskey. At this point in her administrations it had grown dark so by lamplight, Mrs Kunoth pulled up a chair and settled down to stitch. It required in the vicinity of three hundred careful stitches.

The good nurse remained with her patient for days until his fever abated and he was judged well enough to travel by buggy back to Oodnadatta. He recovered from the buggy ride in time go the three day distance by train to Adelaide and hospital. Examined by specialists there he was judged well-recovered and the bush surgery he had received was considered extraordinarily expert.

Louis returned to Oodnadatta and pursued his interest in Lillian, one of Mrs Kunoth's daughters. She was a pretty girl and had assisted in his nursing. He wooed her and won her hand and when they married, their friends included among the wedding presents the offending bullock's horn, safely mounted on a wooden pedestal and inscribed. While convalescing Louis had been following up rumours of a station property for sale near Alice Springs. There were already a large number of horses running on the place and Louis intended to join the increasing Indian Army trade.

Having made a successful bid, he and Lillian set out on their journey to Love's Creek Station and their future, straight after their wedding.

Horses earned them a good living and Louis was one of the staunchest advocates for a 'MacDonnell Range and Stuart' race club. He rode his own race horses and under no circumstances would he miss a meeting

Rationed Rifle Bullets

Bruce Plowman notes in his book

> '....A large part of this area was called bad blackfellow country partly because the Aborigines outnumbered the whites by hundreds to one. Being of fairly high intelligence and better physique, the natives were seen as a potential danger to the white man who travelled alone. They were quick to take advantage of the white man's isolation and to exploit his weaknesses...... North of Alice Springs bushmen seldom go unarmed....between Ryan's Well and Barrow Creek there is always an element of risk.'
>
> Plowman, B. *The Man From Oodnadatta*, 1935, Angus and Robertson

The Glen Maggie family learned while young to use rifles and respect them. Ammunition was expensive and hard to come by and was consequently carefully hoarded. They never felt in any way threatened around their property but were taught to hunt with native spears as well as rifles, and Billy took it upon himself to tutor them. He delighted in the exercise. Bullets were rationed and were not under any circumstances to be wasted, so when young Ben once brought down two kangaroos with one lucky shot, Billy was delighted to be able to announce, 'Well, didden I bin learn im?'

Once when Liz was home alone, a black man approached asking in pidgin English for clothes. Liz explained he would have to work for them and showed him to the axe and the woodheap. She went about her household tasks until some hours later when her woodcutter presented himself and indicated that the task was complete. She was therefore happy to uphold her end of the bargain. Happy that is, until she availed herself of the woodpile and discovered that only half had actually been chopped and piled cunningly to conceal the remainder. Her dander up, she grabbed her rifle and went after him. He had got no further than the well, so she hadn't far to search. 'Give me back that shirt!' she demanded while firing a bullet overhead. In terror her victim took to his heels, peeling off his ill-gotten shirt as he ran. 'Take off those trousers!' This was a difficult operation under the circumstances but she fired another shot high into the heavens and he obliged. Minus his dignity he disappeared, a naked man wearing only a pair of overlarge boots which Liz judged him to have earned with the half-chopped woodpile.

She never told the story for years because she was deeply ashamed that she had almost allowed herself to be outwitted.

Claude, Gene, Ben and Mags

When possible the young brothers took turns away from the property working at other jobs. They were skilled at drilling, droving, fencing, shearing, mustering and branding and were sometimes called upon to act as guides for men pushing further out in search of pastoral land or prospecting for that elusive gold. Required less at home during winter months, they were therefore able to enjoy more exciting opportunities beyond their own boundaries.

The Adelaide University sometimes employed Claude to escort their research teams on bush trips. Ben alternated his work with ventures out beyond known country while Gene, mistrusting his own social skills, was usually the one who kept the home-fires burning.

Having outgrown their childhood, the siblings had become more definite individuals. Claude was tall and slender. His hair was the colour then known in more sophisticated circles as strawberry blonde, but to his family it was simply 'ginger'. He favoured light blue shirts and managed to appear uncreased in any circumstances. He handled horses with the same as did his father and rode as a jockey in every district race meeting he could get to. He was a good conversationalist and smiled easily.

Ben, on the other hand, appeared rumpled. Always hatless, he was deeply tanned and his brown hair had a rusty look. Never neat, always at least 'a month short of a haircut'. Seeming always to have lost his watch, he gave up on timepieces and trusted the sun and instinct to tell him the time. He was seldom hurried. He walked with a slow easy lope and habitually noted everything around him.

Despite owning calendars, the people of the outback were sometimes prone to losing their way with time. Once at Mt Riddock Station, travellers were startled to discover the Webb family sitting down to Christmas dinner. It was a month late. The fact didn't embarrass the Webbs and having unexpected guests added lustre to the festivities.

Gene was the quiet one of the Nicker brothers. He was the peacemaker. The voice of caution. He so seldom took off his hat that when he did, a pale forehead emerged above his eyes. His contemporaries teased him with the question, 'When you undress, Gene, do you take off your hat or your boots first?'

Mags had become wilful; spoiled by her father and brothers, the world was her oyster. Much darker-haired than any of her siblings, it was Mags who had the temper none of her brothers were prey to. They were all reasonable and patient men and perhaps their little sister's fiery temperament had helped develop that aspect of their natures.

Rock Carvings of an Ancient Odyssey

When time permitted, Ben's curiosity led him along the paths of earlier explorers, like Warburton and Ernest Giles. He prowled the Simpson Desert and poked about by camel examining the unmarked areas on his maps. He had no real interest in motors. Their intricacies were better left to other people but he was happy enough to be a passenger. One of his friends told a story that he travelled with a picnic group and they staked the radiator. As none of them were much more qualified than he was, when Ben suggested oatmeal might block the hole, they were more than willing to give it a try. As the engine heated, the oatmeal boiled and exploded peppering both car and occupants.

His map collection overcrowded a worn leather satchel he had stitched together for the purpose and which he took with him wherever he was in case he picked up some information vital to his cause. He carried note books and jotted in them everything of interest which he intended to check out at a later date. Many of his trips were planned to take in points of interest he had heard about as much as to make his own discoveries.

With two good mates, Ben spent a year poking about the western desert generally prospecting for gold and copper. Having read the journals of Alan Davidson's epic prospecting journey of 1900/1901, the three men were inspired to use his route as a jumping off point. As was usual, they spent as much time searching for water and found supplies in great quantities but difficult to access.

They observed a particular type of ant trailing, a group of shrubs a little greener and of a different species, a flight of birds wheeling to an obvious watering point; small birds unable to fly long distances to drink. But when no obvious water appeared they discovered tiny apertures through which only a fist could reach. Rain soaking into these secret catchments is sealed away save for sometimes such minute openings that only by sucking through a stem of hollow grass can a drink be had. Every trip turned up some new aspect of the desert.

A favourite place of Ben's which he kept returning to was the area where three states meet; Western Australia, South Australia and the Northern Territory, Ngaanyatjarra/Pitjantjatjara tribal lands. The country which hides the mystery of Lasseter and keeps many other secrets. The local Luritja language is a master tongue relating to Pitjantjatjara (NT and SA); Ngaanyatjarra (WA) and Pintupi (NT). Having honed his language skills he opened up for himself a more valuable range of interest.

West of Alice in the Cleland Hills is a valley where once the floor must have been much higher than it is today for carved on the rock-face beyond reach, are ancient figures of men and animals which could never have existed here. Characters, symbols and a seven foot, helmeted horizontal human figure, confound general knowledge.

Perhaps centuries before the pyramids were built, a flotilla of ancient ships foundered on Australia's western coast at the mouth of the De Grey River. They may have been irretrievably damaged by storm. Whatever the cause, carved evidence points to a settlement at the mouth of the river, evidently of many year's duration. Probably in the hope of finding another coast, or better food or even signs of other habitation, it is thought that survivors eventually set out in an easterly direction. Evidence suggests that perhaps a stronger more adventurous party set out to blaze the trail which can be detected by carved messages. As the Australian Aborigines had no tools with which to carve rock, it appears obvious that these were a more advanced people. The thousands of etchings they left every fifteen to twenty miles have yet to be deciphered but perhaps they indicated what food was available, where water might be located, what tribal dangers might exist, and which direction to follow. It might be suggested that messages of hope and perhaps sometimes of disaster might have been included. Shown to an Aboriginal elder, his reaction was a shrug of the shoulders.

'That one *itta mucka* (lizard tracks). That one night-time before dream-time.' Whether the Cleland Hills carvings are anything to do with these ancient travellers is debatable.

Geographically, Australia would have been a less arid continent and the crossing perhaps not as daunting as it was for our relatively recent pioneer explorers.

When eventually these enigma are deciphered, we will read a new chapter in Australian history. We will understand something of their struggles for survival as they wandered across a foreign land, through where the weather station, Giles, now exists, diagonally south to the Flinders Ranges in South Australia. Finding no other shore, they turned, prematurely, and directed their steps north through Central Australia. Here they seem to have branched into two separate groups and any further signs have apparently disappeared.

In our desert's many secluded tracts are doubtless secrets yet to be discovered; so easily missed by passers-by within a very short distance.

'Little Bit Long Way Benninick'

For Ben, the possibilities were an endlessly absorbing provocation. He went further and further, widening his inland horizons and filling his notebooks. He carried a sketch-pad and drew wildflowers or shrubs and trees he had never seen before, birds, insects, strange rock formations and extraordinary carvings. These were sent off to an university and probably now lie disregarded in their cellar. The more he found the more he needed to know.

He was so often away somewhere, that when questioned about his whereabouts, almost any Aboriginal informant would thrust out his lips to their furthest extent. Lifting his chin and pointing with it, he would explain, 'Benninick, him bin go little bit long way that way'. Consequently his brothers began to refer to him as 'Little-Bit-Long-Way Ben'.

Ben was inspired by the colours and shapes of his personal Outback, moments he had no way of sharing because, although he carried a camera, there was no real excitement for him in black and white. Within the boundaries of a photograph he could never hope to encompass the full 360 degrees of what he saw and could never capture. He would have liked to paint more elaborately than his little water colour sketches but in his area there were no teachers or access to materials. He knew his limits and knew himself too restless to pin himself down to the learning process.

Born and brought up in the bush, there was an openness in Ben and in his contemporaries. A man was a mate and a man could be himself. Ben was probably more than most, his own man. The seeking, thoughtful, studious

Ben of the bush was another man entirely when he got to town. The customary couple of beers would loosen him up and turn his visit into a celebration. Bushfolk knew one aspect of him; the township, another.

He discarded socks because they gathered and hoarded prickles and shoelaces because they caught in twigs, hats because they blew away and had to be chased. He lost combs and brushes and kept his hair as short as possible but remarkably he went clean-shaven because beards itched. For everyone of his idiosyncrasies there was always good reason, in his mind at least.

He cared for injured animals and visitors learned never to sit on a down-turned box in any camp of Ben's because there was most probably a pet snake underneath. He raised fledgling eaglets to adulthood and thanks to Billy's tutoring could track anything that crawled or ran or hopped. It was said that he could out-track the Aborigines and Billy would agree: 'Well, didden I bin learn im?'

His interest in wildlife, however, took second place to his support for the underdog. At Bob Buck's property, Titra Well Station, Ben was preparing for the Foy expedition into the western desert, led by Kurt Johannsen. Mark Foy was a wealthy Englishman with Australian connections. One of his better-known Australian speculations was the Hydro Majestic Medlow Baths, an elaborate health resort in the Blue Mountains near Sydney. He enjoyed a passionate interest in travel and would frequently gather his family for an interesting safari. This one was in search of any indication of Lasseter's possible whereabouts.

Bob had, like most settlers, free-range fowls including a large number of chickens and was currently doing battle with marauding hawks. Some of the party had been indulging in farewelling themselves when a hawk injudiciously appeared, hovering above them. Bob ran for his rifle while Ben grabbed a spear which happened to be handy. He wasn't very steady on his feet but with spear poised, he reeled in a circle sighting his prey while onlookers expected his backwards impetus to bring him crashing ingloriously to their feet. Apparently defying gravity, the spearman poised and flung his weapon with deadly aim. Onlookers were afterwards to report it as one of the funniest but most impressive pieces of spearmanship they had ever witnessed.

On that same trip, a skilled Aboriginal tracker was travelling with Ben when they crossed the tracks of an Afghan camel team. Having studied the footprints, Ben remarked, 'The third camel from the rear has a sore left hind foot.' The Aboriginal begged to differ. 'No. It's the second camel from the lead.' Another traveller with them noted the exchange and was surprised to discover when they caught up with the team that it was Ben who was correct.

Everybody has an Achilles heel, although not all of us will admit it. In Ben's case, it was thunderstorms which could bring him undone. Only those close to him knew about it, and wondered how he coped when out in the bush alone. It was the noise; the crash and grumble and explosion of violent night storms roaring above and around him he found difficult to handle. Inspired by some incident in his forgotten past, perhaps the storm on the morning he was born, or an unknowing premonition. Who could know. It was strange that warfare should be of such interest to him when sounds closely resembling battle should so unsettle him. Whatever the reason, he handled it and hid it so well that very few knew about it. When a shudder shivered his body too obviously, he would laughingly announce, 'Somebody just walked over my grave!'

He was employed as guide and linesman with government survey teams and enjoyed the work, while gaining a professional qualification in geodetic survey to couple with his natural bushcraft. Summer conditions sent the teams back to their city offices and released Ben to saunter through short-term jobs in the interim.

Sam Irvine, Royal Mailman

Sam Irvine was the Royal Mailman. He took gold shipments to Adelaide from Tennant Creek, and when he was available, Ben rode guard or 'shot-gun'. Although he never had to actually shoot anybody, he enjoyed those trips. It was a thirteen-hundred-mile outback run and perhaps a tribute to the stern front these two big men presented that the shipments were never threatened.

Sam was born in 1890 at Boucaut, in the district of Clare, South Australia. He started work at fourteen on cattle stations and woolsheds and was married at the age of twenty two to the daughter of Tom Farrell who was police officer at Burra.

Mary Irvine was happier in more citified surroundings than Sam was used to. Unable to find work in Adelaide, Sam headed north to take up a mail contract between Kingoonya and Coober Pedy. It was 1920. It must be said that he preferred the outback to city life and the restrictions of marriage and fatherhood. The failed marriage was a burden he seemed to have carried for the rest of his life, but his contribution to the outback was immense. He motorised the mail service which had been previously a camel-train, packhorse system and became a legend in his own time. The roads were not always passable and it was said that if you were determined to get through you should follow Sam in his Reo truck. It was Sam who delineated the difference between being bogged and being stuck. (If you got out in twenty four hours, you were stuck. After twenty four hours you were bogged.)

Becausc he carried hotel supplies, if the bog was really a beauty according to Sam, the severity of it became known as a 'one bottler' or a 'two bottler' and once in a while, if he was really running late, 'a whole b....y case.'

Because hardly anyone had the experience or mechanical knowledge, Sam carried an enormous collection of spare parts and mostly managed to fix anything with a piece of number eight wire. The more hardy vehicles of the time, the Dodge utilities and later the T Model Fords, had long boxes bolted to their running boards in which to carry all the spares they might need but there hardly ever seemed to be the required number of nuts and bolts.

He pioneered the Kingoonya to Coober Pedy run and made his own road by dragging heavy logs behind his truck. He moved on to the Oodnadatta to Alice Springs road and then the Alice to Birdum in the top end. He carried mail, goods and passengers and sometimes when someone was ill, he not only drove them but also nursed them overland to hospital in Adelaide. From one end of the outback to the other, the Inlanders held Sam Irvine in high regard. He knew everybody and kept them all in touch with each other and it can be truly said that he was very much part of the great outback family. His hobby was horse-racing. He raced a horse in Alice Springs, called 'Boxhead' and had one or had a share in a horse trained in Adelaide which ran under the name of 'Barlowerie' and distinguished itself by winning the Grand National Steeplechase in 1930. Sam was always the one to ask before you placed a bet and you were never long in his company before the conversation swung in that direction.

One of the many stories about Sam involved a well-known lady who travelled with him as passenger. On their first night out, Sam made camp and cooked her dinner, then laid out her swag on one side of the Reo and his own a decent distance away. While removing his boots he was startled to find his passenger waving a pearl-handled pistol in his direction and announcing, 'Mr Irvine, I want you to know that this is my protection!' Sam studied her for awhile, stubbed out his cigarette, rolled into his swag and drawled, 'Madam, your face is your protection!'

The Country Nobody Wanted

In 1911 the Northern Territory was transferred from the South Australian Government to the Commonwealth which in turn didn't do very much about it. The Centre battled along without any real help from anywhere. In 1926 things might have altered when the Commonwealth government legislated a division of the Territory at the parallel of 20 degrees south latitude, through the neighbourhood of Tennant Creek. For five years a Government Resident headquartered in Alice Springs and one in Darwin but it made no real difference to the Centre. The economy fluctuated through gold discoveries and the Great Depression. The beef industry stumbled along, voluntarily feeding the hungry locals who were unable to market their cattle. It cost more money to send them to distant markets than the prices they could get for them. When wool prices reached their highest point, a shilling a pound was the best anyone in the Centre could aspire to.

The wealthy 'outback squatter' was a fictional figure. If he was wealthy, he didn't need to live in the outback.

Bush Medicine and Athletics

The pioneers, the adults of this outback world, were all remarkable men and women. Wider skies attracted them. Isolation wasn't any barrier to their aspirations. Together, men, wives and children, erected fences, built their homesteads, mustered cattle and battled the elements; drought, dust-storms and floods. Because they had no other yard stick, the children accepted the lifestyle as normal.

Ben, however, needed to extend his own boundaries where no real boundaries existed.

Prepared for every emergency he carried with him a medicine box made up of a collection of those bits and pieces he found most useful, and he was frequently called upon to succour either an animal or a person. He was naturally aware of many bush remedies and was convinced of the healing power of, among other treatments, witchetty grubs. Mostly found tunnelled in the trunk and roots of the acacia growing all through the outback, witchetty grubs are regarded as a high-protein food source; when squashed into a paste and applied to an open wound or sore or even a burn the effect can be remarkable.

There was an Aborigine called Stumpy who came off second best in a fight at Mt Doreen Station near the Western Australian border. His belly had been ripped open with a knife and a large proportion of his intestines dislodged. When Ben was called, he cleaned it and pushed everything back inside arranged as he thought it should be before stitching up the external wound. Nobody expected the patient to live but he survived to a ripe old age. Because Aborigines had no truck with bandages, a thick coating of mud and ashes better served the purpose. It kept flies and infection from any wound and layered underneath with a witchetty paste, was guaranteed to heal well.

When once his camp was rifled and some of his possessions stolen, he scouted around his camp and tracked the miscreants down. He then mixed up two good strong doses of Epsom Salts and made them drink it. The result was so discomposing and undignified that word quickly spread and his possessions were ever afterwards secure.

Mythology and ancient history continued to intrigue him and he carried books in his saddle-bags to study whenever and wherever the opportunity arose. If he had any regrets it was that he had been a child and had missed the chance of serving in the Great War. So many of the men he met had been a part of it. In fact the Northern Territory had seen 40% of its male population off to fight overseas. It hadn't been easy for any of them to even reach a recruitment centre. They had travelled by foot overland to Brisbane, or by camel and horseback, even bicycle to Adelaide.

Ben enjoyed his solitary sojourns but was equally appreciative of company. He enjoyed his trips to town. The little place clustered around the foot of Anzac Hill, and there was a night when Ben led his friends, all as primed as he was, to the summit whence they serenaded the unenthusiastic population. It wasn't a popular group for a day or two, but Ben's persuasively friendly personality was such that the incident was quickly forgiven and even embroidered to more dramatic proportions. When sacramental wine was replaced with whisky in the newly erected church, every finger pointed to Ben.

He played tennis and cricket and because he was fit and fleet of foot, no trip to town would be complete without someone seeking the opportunity to challenge him to a foot-race. Always begun from the hitching rail in front of the Stuart Arms Hotel, it was the publican's role to act as starter and hold the bets. Ben was never beaten but there was a draw when one of his survey team challenged him. Circumstances moved both men beyond the opportunity to ever re-contest the race.

Sam encouraged social gatherings when and where the opportunity arose and one of his achievements was the annual Christmas Sports Day event. The athletic prowess of the population was, not surprisingly, outstanding because they had to be healthy, sturdy and indomitable. Among the usual competitors was an Aboriginal called George King whose cricketing skills were legendary. It was said he could throw a cricket ball at least fifteen yards further than anyone else. He was well-liked and when he died of fever on a droving trip further north, the Centre mourned his untimely end.

Competition was fierce and there was often more energy than finesse.

At one of these meetings, two mules in the process of being harnessed to a buck-board were exhibiting more than the usual mulish obstinacy. An audience gathered with unhelpful comments and useless suggestions only succeeding in confusing the predicament to a greater extent. Finally, watching in exasperation, Charlie Dubois and Harry Kunoth, both powerful local men, stepped forward and picked up an animal each, bodily lifting them into the shafts.

Without machinery such super-human strength was needed. Often, in isolation, previously untested reserves had to be called into action but in this situation a few drinks and high spirits were enough to fuel the special effort.

The Kunoth family were of German origin and when one of the second generation departed for the First World War, his mother asked if it was right that he might have to shoot one of his cousins. The soldier replied, 'Well Mum, I'll aim first and then shut my eyes to shoot.'

Michael Terry the Explorer: 1928

In 1928 the Glen Maggie homestead, quietly minding its own business and enjoying another quiet day, thought they heard a distant drone. The sound swelled and out of a cloud of dust appeared two six-wheeled Morris Trucks. Sheep and cattle bolted in terror. The station dogs, uncertain whether to retreat or attack, decided on the coward's way out and went for cover behind the fowl house adjacent to the outback dunny. It was the Michael Terry expedition between Port Hedland on the north Western Australian coast, by way of Tanami, Alice Springs and Adelaide to Melbourne.

Michael, an Englishman, had served with the Royal Naval Air Service Armoured Car Unit in the First World War. Taken prisoner in Russia he had been involved in a successful mass escape from Bolshevik forces and struggled across Russia by subterfuge and cattle-train. Eventually arriving in far northern Murmansk, the escapees managed to make contact with a British naval ship and were returned to England and medical facilities. Many of the men were suffering from gas inhalation in addition to general debilitation and as a result, were invalided out of the army. On medical recommendation, a dryer climate seemed a good idea and Australia attracted Michael's adventurous inclinations.

With his military-trained mechanical background, he found work in outback garages but it was Australian distances which attracted him and he set about fund-raising through British contacts. With fellow countryman Richard Yockney, he crossed from Winton in Queensland to Broome by car. No one had previously attempted such a trip. They had nearly perished in the attempt but undaunted had then taken on a second venture against good advice, from Darwin to Broome. With British acclaim still ringing in his ears, Michael was delighted to accept the loan of two worthy vehicles from Sir William Morris himself for his third crossing of the Australian inland. It was this trip which brought them to Glen Maggie.

The party had not long arrived when Ben rode in with camels and the stage was set for a mutual admiration which was to last a lifetime.

Watching the team striding towards the well, Michael's mind was instantly set upon possibilities of exploration with camels, and Ben became his focus. But he had other commitments to fulfil and it would be awhile before his ideas could be put into practice.

These were depression years and gold and minerals were thought to be out there waiting to be discovered, therefore any one who had a few pounds to spare was more than willing to invest in Michael Terry's ensuing prospecting expeditions.

Thirty seven years later Michael Terry wrote:

> 'Self-educated Ben was yet the best educated man I ever met. He was a voracious reader and always an interesting talker.....when I first met him I knew at a glance here was a man who would quietly tell you to go to hell if you tried to order him about..........His deep-set eyes looked straight into you, not aggressively but with keen perception. Actually he was a fun-loving man with an easy-going temperament, calm in danger, anything but calm at play.'
>
> Michael Terry, Sept. 8th, 1965 *People Magazine.*
> 'Portrait of a real bushman.'

Mags Drives Overland from Adelaide to Darwin: 1929

Meanwhile Ben fed his journals and prodigious memory with co-ordinates and details of his bush trips, and his bank balance with seasonal work. The Nicker children had become adults. In 1930, two years younger than Ben, Mags was twenty. She had completed her education in Melbourne as had been previously arranged on that long-ago trip south, and had enjoyed it. When she left home she had been a very anxious child. It had been a few years and in the interim she feared her uncle and aunt would never recognise her. She arrived wearing handmade clothes and because Liz had no way of keeping up with fashion trends and fabrics, the wrong hat for the fashionable Melbourne child, and boots, when everyone else on the station platform was smartly shod in shoes. Her hair hung in plaits. Everyone else wore bob-cuts. Her complexion was deeply tanned. She need not have worried for she stood out as an obvious child of the bush.

Her young, feminine heart was so offended by her predicament that never again would she allow herself to be embarrassed by ignorance of fashion. She revelled in the competition of classroom study and enjoyed her time in Melbourne. She accepted life as it came and made the best of it which was just as well because she was not easily able to return should she have become homesick. From Melbourne to Adelaide was then a slow trip by train, then it was another three days to the railhead at Oodnadatta. Before cars it took ten days by buggy and horses to Alice Springs and a further two or three to Ryan's Well. Under these circumstances it isn't surprising that Liz visited her only once and her sister, Anne Jane managed one trip.

She never adapted to the city, rather she learned to underline her background and establish a kind of distinction in her own small way.

Leaving school she settled down to nursing. Admiring her mother's self-taught nursing skills, Mags realised that training would equip her for whatever direction her life might take. She enjoyed the work so much that she seriously considered giving up nursing in favour of studying medicine, but news of her father's ill-health intervened.

While preparing to take leave, she was asked to drive a Wily's Overland Tourer to Darwin in the company of Lottie Mackay, one of her southern cousins. Six years older than herself and without driving skills or knowledge of the outback nevertheless Lottie proved an excellent companion. Their pre-arranged agreement was that Mags would do the driving and Lottie would change the tyres. Her poor cousin had the rough end of the stick as they blew a lot of tyres.

Finding Sam's health seemed fairly stable when they reached Glen Maggie, the girls journeyed on with lighter hearts and the confidence of having already negotiated 1700 miles. The road wasn't much more than two wheel-tracks between telegraph poles and they still faced the problem of creek crossings but they were only stopped in their run to Darwin by an earlier-than-expected wet season which flooded tropical river-crossings and rendered them impassable.

Once when they pulled into a Telegraph Station, they found a telegraph operator who lay obviously critically ill. Having achieved what she was later to describe as an over-abundant self confidence, and some measure of nursing skills, Mags took over. Taking all the patient's vital signs into consideration she had his fellow officers send off a telegram by pedal radio to Dr Clyde Fenton in Darwin. His reply confounded everyone for it read 'Recommend A.P.C. four hourly.' None among the company could offer any clues until Mags had the bright idea that it had to mean ...'Advise Patients Condition', which ritual they performed conscientiously every four hours. Mags was horrified when she later learned there was a new wonder-analgesic called A.P.C.

> 'Early in November, 1929, two girls passed through Katherine, North Australia, in a car. With a badly knocking engine, three leaves broken in the front spring and minus headlights, and the wet season of the tropics steadily setting in, no-one experienced in the country could imagine that they would get through. To a warning given came the cheery reply, I guess we can push the old bus through. They did, and while little publicity was given to the feat in southern areas, yet they made history along the overland route across Australia for a feat which necessitated tremendous courage,

self-reliance, and will power to accomplish in the face of so much adversity.'

'Out Among the People', *The Advertiser,* May 1931

Sam Suffers a Stroke: 1930

Having been away for some time on one of his ventures, Ben was not aware that Sam had suffered a stroke. There hadn't been any way of getting a message to him. Liz, knowing her skills inadequate in this case, nursed him with assistance by telephone from the Australian Inland Mission. He seemed stable and on the mend but another seizure impaired him and when he was able to sit comfortably in a wheel chair she bundled him up and with Eugene driving their newly acquired motor, took him to hospital in Adelaide.

This time they drove the two hundred odd miles between settlements and were able to put Sam under a roof and between clean sheets every night. By the time they pulled into the portals of the Royal Adelaide Hospital, it was Gene who was in need of aid because the city and its traffic had been an unforeseen ordeal. In 1930 he would confidently drive across a desert but city streets were another matter. For Liz, it was a relief to give Sam into the care of the medical profession and to have brought him the distance without incident. For some months she spent most of her days by his bedside. By the beginning of December, Sam was feeling better and sent Liz home to Glen Maggie to be with the family gathered for Christmas.

Sadly, a telegram brought news of his death on Boxing Day.

Alice Springs Grows Up

By the 1930s, even Central Australia was not locked away from the march of progress. The family had struggled into the Centre by horse-drawn buggy and could now travel more comfortably by motor car or train. They had witnessed air travel; telephone and radio kept them in touch with a world which was no longer so distant but a part of their lives. In Alice there were two buildings, a church and a dance hall built on stilts and entered by stairs. In themselves, stairways represented a sort of sophistication. Alice Springs was going up in the world.

It was still a rugged little settlement where dirt roads divided around standing gum trees and bogs lay in wait for the unwary when it rained. Strangers from distant places visited and sometimes stayed. An Anglo/ Indian moved about the township treading with such languid dignity that it seemed an invisible entourage of servants accompanied him. He lived in a cave part way up the western slope of Mount Blatherskite. Two Aboriginal women attended to his clothes and meals but otherwise he didn't encourage companionship. He was such a man of mystery that only his regular remittance cheques identified his name as Drummond-Hay. When he disappeared as suddenly as he had come, the riddle of who he might have been, why he had appeared in Alice and what had happened to him was never solved.

The Australian Inland Mission building in the centre of town, stumbled along for a number of years. Its erection was frequently halted by the difficulties of bringing up material from the railhead at Oodnadatta. Bits came initially by camel team. Then Wilkinson's store at 'Oodna' started a motor service, bringing supplies to their Alice Springs depot. When space and travelling conditions allowed, they lashed on board what roofing and timber they could safely carry. Builders wandered on to other jobs or left the district. But in 1926 the most sophisticated building in Alice Springs, the Australian Inland Mission Nursing Home, was finally opened. Alfred Traeger installed his newly invented two-way pedal radio in a small stone building adjacent and this distant township took a giant leap forward. Two qualified nursing sisters were permanently installed and no longer was Liz required to tend the sick. No longer did the operator at the Overland Telegraph Station have to relay medical assistance from a doctor in far-a-way Adelaide. John Flynn's 'mantle of safety' was a reality.

Claude Nicker

Claude had moved to a nearby property north of Glen Maggie, called Woola Downs and was getting himself firmly established when his father died. Claude was the more quietly dignified of the three brothers. Tall, ginger haired and angularly slim, he was a welcome guest and companionable host. He arrived always bearing gifts, of homegrown fruit and vegetables, fresh-baked bread and station beef and mutton. His library was painstakingly collected against the day when he expected time to spare but it never happened often enough and he was always 'three bookshelves behind'.

Like most bushmen, his hands were never still. He was clever at plaiting and all the associated leather crafts; he liked to carve and create. The household furniture bore witness to his talents and his garden provided fruit and vegetables in most seasons as well as shaded pergolas and flower beds. He managed to provide an hospitable oasis of home-made comforts.

He was as much a bushman as his siblings but when he became lord and master of his own domain running an equal mix of sheep and cattle, his time away was limited. His upbringing had prepared him for any eventuality. It was all second nature to him and he was relaxed and capable whatever the circumstances.

Aboriginal dialects came easily to him and it was commonly acknowledged that he was proficient in at least a dozen of them, a talent he had honed while droving and working throughout the bush. As a white man he was better able to converse tribally over a wider area than Aborigines who were limited by divided cultures and tribal boundaries.

When he married, his bride found the reality of isolation unnerving and within a year the birth of a still-born son snapped irrevocably the tenuous threads holding them together. She returned to her former lifestyle in Melbourne. The infant, never having breathed, was buried on the isolated outskirts of Alice Springs where today a suburb sprawls over the grave.

Liz Moves into Alice Springs

With Sam's death, the bereft family considered their options and agreed to put 'Glen Maggie' on the market. Liz elected to retire to Alice Springs where she would be nearer to Anne-Jane and her grandchildren at 'Undoolya', a mere ten miles out of town. She could involve herself in the gardening she had been denied at 'Glen Maggie', and enjoy the social life of the Alice.

Suddenly she was able to buy her bread and meat at Lackman's store. She could chat with other women over cups of tea and cake. She could take a basket and go shopping and attend church service on Sunday. Ice was delivered when she wanted it and Mick Heenan brought fruit and vegetables to her door. Everything to do with town living was a joy to her and a new experience. With Claude at Woola Downs and Gene interested in a property south of town called Todd River, Ben was involved in his own 'knocking about' lifestyle and it had nothing to do with settling down in any one place.

Thinking about Sam, the sale of Glen Maggie and her new circumstances, Liz went to bed one night and dreamed that approaching the cemetery where Sam was buried she found him waiting for her beside the gate. He assured her that it was Mags who now needed her attention.

The deeply spiritual nature and isolation of the country in which they lived, must have honed their perceptions, for premonitions and dreams seemed to play a role in the lives of many outback folk.

With all the bustle of moving and settling in, Liz could have done with more than perfunctory support from Mags, but Mags had a personal problem. She was pregnant and unmarried. As far-flung as the bush was geographically, it was a close-knit community and Mag's condition wouldn't go unnoticed indefinitely.

She retreated to her sister at Undoolya and received first the scolding then the support she needed. Anne-Jane brought mother and daughter together. Understanding her dream of Sam and the message he had given her, Liz felt herself forewarned and therefore forearmed.

When the infant arrived, it was thought that Mags was coping well but three months later when shame overwhelmed her and suicide seemed preferable, mother and child were pulled out of a waterhole. Unknown to her, a twelve-year-old nephew was passing over the shoulder of a nearby hill and saw what was happening. It was fortunate that he had gone to alert the goat-shepherds to bring the herd in a little earlier that night. While he dived into the water and dragged mother and child to safety, a shepherd ran to the homestead for help. Afterwards he was heard to say,

'I didn't know I could swim.'

And many years later when asked if he swam, he replied,

'I did once.'

There were not many in the outback who could swim, as water-holes were too hard to find and drinking water too precious.

Mags' brothers, shocked at her despair, tendered comfort typical of them. Ben announced, 'I'm an unmarried uncle!' Claude was more gentle, so shortly after his own bereavement, 'She's a beautiful baby and you're a very lucky mum.' Gene failed to understand the fuss, only commenting after studying his infant niece, 'If you want a baby-sitter any time...'

It was obvious everyone else came to terms with the situation more easily than Mags did. Life settled down.

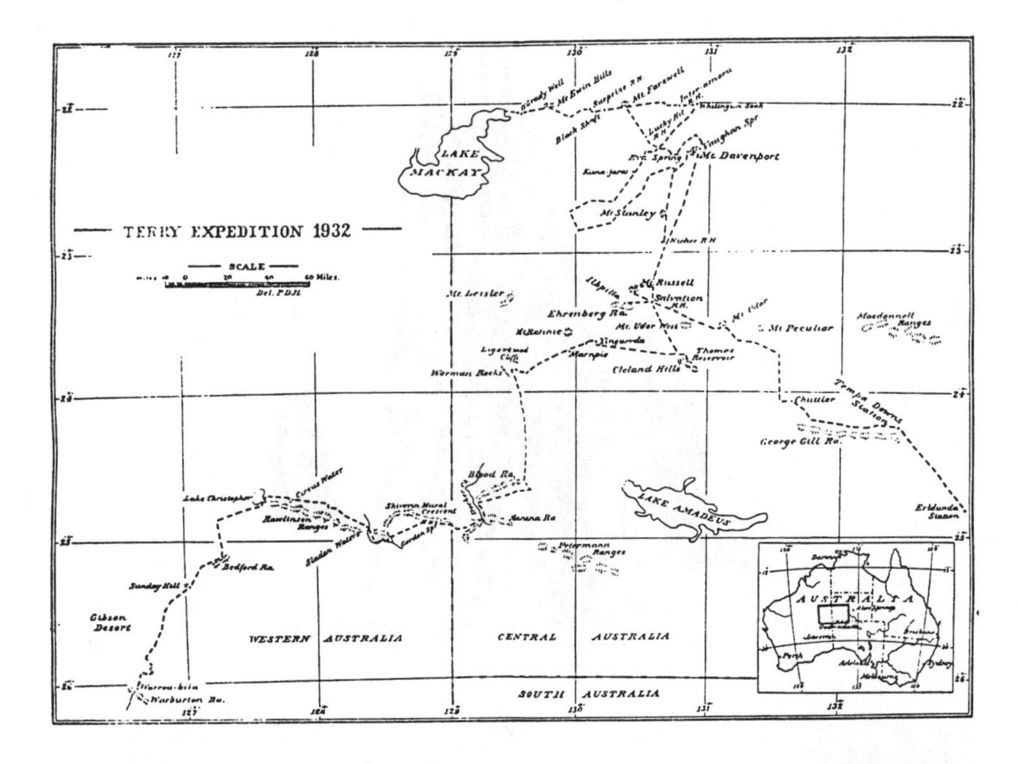

Piece of rock carving at Cleland Hills, NT.

4
Outback Expedition
1932

From Erldunda to Leonora

On June the 26th,1932, leading a string of camels, Ben crested a sand dune and walked into Erldunda Station where Michael Terry, Stan O'Grady and two Aboriginal camel handlers, Lockie and Jack, awaited him. They were prepared with all the stores and equipment considered necessary for many months away, while Ben supplied saddles, canteens, camels and knowledge of many areas through which they planned to travel.

Of the twelve camels, three were half-broken riding mounts and the other nine were pack. Two of the strongest carried a pair of water canteens with a capacity of twenty seven gallons each.

In every camel string, one stands out as the leader and in this case it was Darkie whose loading of foodstuff at the trip's beginning was roughly eight cwt. Divided among the team were all the other necessary adjuncts and bits and pieces. Loading was simplified by the process of laying out packs and boxes with weight adjustments in two long parallel lines and walking the camels between, then hooshing them down.

If any camel was unused to the process, or just plain ornery, it had to undergo the indignity of a tie down until it was safely loaded, which was a difficult process requiring a great deal of both strength and patience. The camels didn't like it either.

They emitted a chorus of grumbles and belches, groans and complaints and if their handler carelessly worked within range, a few nips and shoves or even a slobber of foul-smelling regurgitation was small revenge.

Sometimes they were known to go one better and roll with half a load almost secure, then all the groans and complaints weren't coming from the camels. They were experts in one-upmanship. A few days out and every camel's temperament was obvious. The nervous, the tantrum-thrower, the sly, the plodder and the one or two who quickly settle in to the routine with a 'let's get this over with' attitude. There is also the lazy grumbler and the joker in the pack.

Ben was the lead. He would take up the lead nose-string and step out at a steady walking pace which would average them roughly one hundred miles a week, taking into account frequent stops for load adjustments, broken nose-lines or any other unforeseen event. He sometimes rode but his preference was to 'footwalk'. If any of their string threw a temper tantrum, his remedy was to double load the miscreant for a few days until it settled down. There wasn't a camel-attitude he hadn't dealt with before, some other time, some other trip, which put him one step ahead of whatever tricks they might like to try.

From Erldunda they travelled to Henbury and on to Titra Well where one of the camels went lame when it staked its leg. Ben dug out stray splinters, covered the wound liberally with tar and sewed a clean piece of hessian bag around the wound. It was a good enough excuse to rest the team.

When a Camel Gets the Hump

When fat camels start out, they quickly lose condition before the muscles build back up and after about a week out it is time to adjust, overhaul and mend. Saddle pads are re-stuffed with dry grass to better fit new contours and patched where cloth has rubbed and thinned. Ropes are double checked and respliced. Every consideration is attended to for the camels' comfort and security, remembering that they carried no radio and the camels were their only life-line.

They moved on across a winter hinterland where softer sunshine coloured the landscapes in gentler hues but still the fierce spinifex stabbed at the unwary. Ben, in self-defence, cut and sewed a pair of leggings from a goat-skin he had acquired from Bob Buck at Titra Well. It was a billy-goat skin and a bit 'high' but weighed in the balance, and despite his team-mates' ribbing, his leggings served their purpose well. He was the one doing most of the walking and his fellow travellers simply stayed off his windward side.

This particular trip was begun in winter. Centralian winter days are glorious. The nights can be freezing but better walking weather can scarcely be found anywhere. The party planned and hoped to have overcome much of the distance before summer caught up with them. There were two major considerations to take into account. The season was a dry one which meant that water would be scarce. On the other hand camels would make little or no progress in mud. They had to gamble on sufficient rains to top up supplies.

At the George Gill Range they made camp by a creek a little distance away from the Tempe Downs settlement owned by a young Bryan Bowman. There were a few bush camels in the yard, unhandled and unhappy in this strange new element and one of them in particular caught Ben's eye. If he could ride it, he would consider buying it. He managed to saddle, halter and mount it and get beyond the confines of the stockyard before all hell broke loose while he clung to the saddle with a do-or-die determination. Finally the girth broke. Ben came as close as he ever had to flying and walked in a most peculiar fashion until his bruised rump healed.

It was fortunate for him that on leaving Tempe Downs their path was across the Nineteen Mile Valley where walking was straight forward and easy. Thirty limping miles later, when they reached Oolcutta rockhole, he was able to soothe his dignity and dwindling pain in cold, clean water.

Gathering Data

Every day the team had other duties to perform. For the Adelaide Weather Bureau, they registered wind strengths, directions and cloud formations. If rain fell, which it didn't do very often, they measured it.

They drove a posthole auger three feet into the soil and took core samples.

Lateritic gravel delineations, herbage and grasses, scrub lands and general land levels were noted on behalf of the Waite Agricultural Research Institute, also in Adelaide. They settled down seriously with protractors and dividers, with base-line triangles and side bearings and drew sketches of topography, calculated distances and registered their bearings. But their first priority was the search for gold. Their office-box was most assiduously protected and almost always in use.

Finding water in a rockhole at Oolcutta, they indulged in a break before weaving their way across an undulating landscape towards Mount Peculiar, running through a mine-field of lethal emu bush. Roughly thirty leaves will kill a camel. They chopped down the lower branches of Kurrajong trees

Marked tree, Yintardamarru Rockhole. Ben cut the date 11.8.32.

Ben with eaglets.

Ben mounted, watching preparations for departure from Mt Doreen.

Ben unpacks at the end of a desert trip.

Ben, Lockie and Stan packing camels during the 1932 trip into the Western Desert.

Nicker Creek. Murraba Ranges, Great Sandy Desert.
Photographs by Bill and Bernie Manley.

Ben carving tree during 1932 Terry expedition.

Ben beside Bonny, his riding camel

Foy expedition camped at Docker River, Petermann Ranges with Ben's camels loaded in the foregound. Photograph from collection of H.V. Foy.

fortunately growing there, appetising stuff for camels, and they fed them until the animals were satiated and had no room left for an investigation of emu bush.

An antidote, if caught quickly enough, is a large dose of Condy's Crystal solution which oxidises the poison. Old-timers swore by a beer-bottle-full and it was suggested that by the time the camel handler had himself put away a full bottle of beer, the camel would be well on the way to recovery. Emu bush, known as *immorninga* among the natives of that area, is a nicotine narcotic. Thrown in a waterhole, the Aborigines used it to stun emus so that they were easily caught. A very small amount will kill cattle. There is a less lethal form of it known as *pituri* and the dried leaves are crushed and mixed with the ashes of other known plants. This combination induces a state of euphoria especially useful in childbirth and initiation ceremonies. It is addictive and often evident carried in a wad on a protruding lower lip.

Veering away from their route towards Mount Peculiar, they turned in a westerly direction to Mount Udor West. Like an apostrophe at the end of foothills of the MacDonnell Ranges it rises in high-domed quartzite prominence, an excellent vantage point; and having climbed to its peak, they were able to sight clearly the mountains Liebig, Russell and Lyell-Brown. It was towards this last they directed their march.

Kangaroos and scrub turkeys abounded and the change of diet stimulated their tastebuds. Ben and his brothers had a penchant for curry and outdid each other with variations. Anyone travelling with Ben knew what to expect when his turn came for culinary duties. They carried salt beef, rice, raisins, dried fruits, tinned foods and shot game when necessary. It was said that if there was no curry powder, Ben would starve. He believed there wasn't any such thing as a bush tucker which didn't cook up into a good curry. Mostly it was so hot one never knew what the basis of it might have been, or dared to ask.

In an Oasis at Vaughan Springs

Ilbpilla, the Mount Lyell-Brown rockhole, proved dry. They had fifteen gallons of water in their canteens and it was a hundred miles back to their surest supply at Haast Bluff. A shower of rain fortunately overtook them and left several hundred gallons of clean rainwater in a wide shallow rockhole. Somebody had to be looking after them.

They watered the camels and hobbled them out to graze before giving vent to their elation and discarding their clothes and inhibitions. Bathed and naked, they laundered everything and settled down to light their pipes and break out a little medicinal whisky by way of celebration. It had been a month between baths.

They named the place, fittingly, Salvation Waterhole which was as near as they came to making a verbal confession of their individual anxieties. Not now having to face the two weeks lost time in returning to Haast Bluff, they directed themselves toward Mount Davenport and travelled light-heartedly through afternoon showers of rain which they could watch playing across the landscape. It was a wide world with no seeming boundaries except for the blue-wrinkled ribbon of low hills ahead and Mounts Stanley and Cockburn both standing like lonely amethysts against the horizon.

There was no wildlife here. Nothing breathed. Only the wind and the rain beside themselves moved with any intent and a wedge-tailed eagle regarded them from its free-floating vantage point.

At Mount Davenport they located Vaughan Springs, one of thirty surrounding the base of this remarkable quartzite hill. It is far from sufficiently high to be truly a mountain but it is shaped on top like a great concave basin where rainfall collects and seeps through subterranean layers to gurgle out in the multiple springs hemming its baseline. The seasons come and go, winter, summer, good times and drought and still most of the springs splash cool, clear, life-giving water. Nearly all inland bird-life is represented here and wildlife is abundant. It was a happy stepping stone in their trek across the desert.

Stan was on cooking detail and discovering a tin of dates at the very bottom of the box, he threw together a date duff. While it cooked, he carefully planted the seeds in secure pockets around the edge of Vaughan Springs and the three men amiably considered that after the seven years required for it to grow and bear fruit, Ben would be the most likely to return and reap the results.

This was one of Ben's favourite camping places. He pointed out the strange lack of anything Aboriginal. No signs of footprints, wurlies or camp-fires, and considering the abundance of water and berries, it was a special place deserving of every respect. They could stay no longer than was necessary.

In Warramulla Country

They packed and moved around the northern flanks of the mountain before heading west, pushing through thickly-growing mulga interspersed with granite outcrops. They were now in Warramulla country. In the local Aboriginal tongue, 'warramulla' is the word meaning 'warrior'. They knew the country could be hostile and were prepared to travel circumspectly, anxious not to cause offence. Knowing their every move was noted they hoped the very mass of their twelve camels, strange and awesome beasts, were in themselves a protection. There is a story told that once, on first coming across camel tracks, a little group of Aborigines studied the evidence very carefully before pronouncing that here indeed was a very strange thing. Obviously the tracks had been made by a group of babies forced to travel by hopping along on their bare bottoms.

As a double insurance, they set light to isolated patches of spinifex as they travelled, aware that the columns of smoke weaving skywards were in themselves an announcement of their movements. 'We aren't creeping up on you. We're passing through with no ill-intent.' It was a politeness.

It was Ben's job to guide towards whatever destination Michael selected and when the distance was made unclear by scrub or ridges, he took his direction from shadows and an infallible sense of direction.

Smoke talk, or signal fires kept them alert and they found a camp where a meal was left half consumed. Whoever it was had obviously left in a hurry. It took two days to track down the people and question them. With Ben's background in dialects together with the more universal finger talk he was able to discover that there was now no water at Lake MacKay, where they were headed but there was plenty to be had near Mount Singleton at what his informant referred to as 'water-snake place'. The Aborigines accompanied them for a couple of days and as communication became more fluent between them, they discovered that these were people they had heard about, but hadn't believed in. They were the 'no water people'.

They were trained from birth to survive by substituting the juices of native foods like yams, roots and fruit. Before Ben could discover more than the most pertinent information and converse more amiably with them, they had, without warning, melted into the scrub which was so thick around them that it was difficult to push through.

Mount Nicker and the Lucky Hit Rockhole

One of the camels suffered a bad stake in his foot and a halt was forced so that it could be attended to. Ben fashioned and shaped a boot out of soaked, softened greenhide to ease walking and protect the wound. Taking the pace more slowly when they moved off, Ben scouted on ahead locating the soak they had been pointed to. There wasn't a great supply but it served their purpose. This one was within sight of a low-lying range, the highest point of which Michael decreed should henceforth be known as Mount Nicker. Directing their route back towards Mount Singleton, across stony mulga, they came out near sunset on a claypan which offered perfect conditions for a night camp. Animal tracks led toward a granite whaleback outcrop nearby which in a long narrow crevice they found a good water supply and were able to replenish their canteens. So pleasantly surprising was this discovery that they gave it the name, Lucky Hit Rockhole.

A fresh Aboriginal footprint warned them that they had other company. That night they spread their packs and boxes and laid their swags out between them so that in the event of possible attack they had some small protection. They ate early and covered their fire so it would not act as a beacon and strangely, having done all that, they slept soundly. Next morning they found new footprints within feet of their camp and it was obviously curiosity rather than animosity which surrounded them.

At every sparse watering stop they made certain to take only what they required, ever conscious of the fact that others needed it as much as they did and they were reluctant to cause any unnecessary antagonism. But the Aborigines who appeared to them from this point on were not helpful. This was an area not known to Ben. They found themselves directed from one foul, black, unusable supply to another. They were, in effect, intruders.

Although aware of the excellent water supply at Mount Singleton, they would avoid going there if possible due to its obvious Aboriginal significance. With canteens filled at Lucky Hit, they were confident of reaching Lake McKay so turning back to their westerly bearing, they plodded across featureless desert entirely ringed by flat horizons and traversed occasionally by sand-hills, up which the camels struggled and down which a slow descent was necessarily forced to prevent nose-lines from breaking.

From the top of every rise they peered into the distance searching for the outline of the McEwin Hills. Much of what they sought was hear-say. No mapping had been done in these regions and the co-ordinates of earlier explorers could not be entirely depended upon. Sketch maps sometimes pointed casually to 'a low hill' or 'a native well', but the water supplies suggested were entirely dependent on rainfall.

A Mysterious Well

Sometimes the team took turns riding out on scouting forays and it was Stan's misfortune here to come across an elderly native woman left to die. She sat emaciated and immobile within the remains of an abandoned camp and were it not for the faintest humming sounds issuing from her parted lips, Stan would have considered her already dead. Around her lay her few pathetic possessions but no food or water. It was difficult to resist giving her a drink but he had himself been a bushman long enough to know that to do so would prolong her life and defeat the purpose of her situation. She was obviously too old, too crippled, too big a burden on her tribe and death was the final gift she gave her family and herself. Aboriginal custom, where transport and medical attention were unknown, was that this would be her conclusion and she would be ritually farewelled, drugged and left to carry herself into the arms of the hereafter. To ease her journey and announce her arrival in the other world she intoned her death chant.

Stan rejoined the others a shaken man because this was a situation he knew about but had never actually witnessed. To have interfered would accomplish nothing and might well have endangered themselves. It was a burden he carried with him, mulling over in his mind and discussing what he might have done and struggling to accept his powerlessness. To comfort him, Ben suggested that it could well be a more refined way to end a long and useful life than white man's culture offered. It was food for thought.

Day followed day.

Finding at last the hills they sought, at the eastern end, within a place they called Sandford Cliffs in the McEwin's, they came accidentally upon a mysterious well which could only have been fashioned many centuries ago by a race of people far advanced in tools and technique.

Because Stan had been the one to literally stumble upon it, they named it O'Grady's Well. On first sight it was no more than a shallow debris-filled indentation but because his thirsty camel appeared more than ordinarily interested, he was led to scratch away a little before alerting the others. As they dug they discovered a slanting, spiralled chute hewn through thirty feet of sandstone and designed in such a fashion as to preserve the water supply from dehydration. When they reached it, alas, it was rancid and unusable. But the mystery remained and does to this day. There are many theories but no real answers. Do any or all of these desert puzzles connect with each other?

Saved by Budgerigars

The team pushed west a further ten miles to the shore of Lake McKay, through samphire flats and dwarf ti-trees. They dug where soaks were indicated but the barest bit of moisture they were able to uncover was almost pure salt. The camels had no purchase on the salt-lake surface and were so thirsty they had to give up any ideas of crossing it, and retreat. There was a distant range westward where Terry thought it was probable they might find water. Ben voted against the idea. Stan felt more positive but Michael weighed both arguments and favoured Ben's reasoning. Ben was the more seasoned bushman. They turned back towards Mt Singleton.

Thirst drove them day and night. Dehydration clawed at Michael, both mentally and physically but he was heartened when he looked up from his camel-perch and saw the line stretched out ahead of him, packs intact, rhythmically plodding forward. But could they make another day?

Travelling fast, the camels bellowed and whined and a couple of them threw fractious fits. Clouds built up and a muggy heat taunted them, but no rain fell. When they rested, the animals refused to eat. They were too thirsty. Back to nothing at Sandford Cliffs. Back to old disregarded wells. Back to dry Blackstone and waterless Mount Farewell. They kept moving, resting only when they had to. Rain built up daily, promising but not delivering any more than a few reluctant spits. Although originally disinclined to go there, now their need to water at Mount Singleton was desperate. The men's limbs grew painfully heavy and their camels tripped and stumbled but kept plodding.

Sometimes in exasperation, they watched a very distant shower beyond any help to them. Finally, within sight of their goal, they were led by budgerigars to Inta-amoru, a rock-hole difficult to access and watched over by four obviously ill tribesmen. By a slow process of billycan-to-bucket-to-camel each beast gulped enough to ease the worst pangs of thirst. The water was foul and smelled but they boiled sufficient for their own immediate requirements and flavoured it with strong coffee. They moved on next day across northern foothills to Whittington Soak on the eastern flank, where a better and safer supply gave them the chance to re-hydrate the camels and themselves. For sixteen days and over two hundred and fifty miles they had battled the odds and almost lost. At a later date they discovered that the beckoning range they had sighted from Lake MacKay harboured soaks which had indeed been dry and had they not turned back they would certainly have perished.

Eaters of Stone?

With the first official day of spring, September the 1st, a refreshed party headed out on a south-westerly course intending this time to round Lake MacKay by its southern shore. An abundance of low rock surfaces motivated a fever of sample-dollying and their odd actions brought lurking Aborigines out into the open to better examine such peculiar goings on. Surely these pale people weren't eaters of stone? During the process of question and answer it was suggested that if they gave up this ugly habit they might hope to gain a more healthy skin.

Another day and a deeply-worn footpad led them to the discovery of Kuna-jarai, a soak at the bottom of a granite hill. Nearby, a stark white ribbon of stone wound down the granite cliff and two white, rounded rocks at its base gave the place its name. It was very obviously a dreaming place of the great snake and the two white rocks were evidence of its bodily waste. This was evidently the place they had previously been told about and when a small band of friendly Aborigines materialised they enthusiastically related the mythological story.

Put more bluntly it was the place where, in the Dreamtime, the sacred serpent stopped for a toilet break.

They camped that night a polite distance from the water and considered their options. For two months they had searched the desert more for water than the gold which was mainly their aim.

These Foolish Men

They had advanced and retreated and despite hovering thunderstorms, the drought had still not broken. They could turn back or go on. Unwilling to accept defeat, they agreed to make one final effort to cross the Western Australian border using Thomas Reservoir in the Cleland Hills as a jumping-off point.

Followed by dry thunderstorms, they headed south, step by step, sweating in the humidity and dust, a combination resulting in sticky, muddy skin, sweat blisters and heat rashes. Summer had come in early. To lift their spirits they sang, a sound which initially startled the camels, but in time they resigned themselves and despite their more discerning natures, stepped out in rhythm. Ben introduced Rudyard Kipling's Jungle Book 'Camel Song' from the 'Parade Song of the Camp Animals' and it was quickly taken up by the others.....'We haven't a camelty tune of our own, to help us trollop along, but every neck is a hair trombone.....' Not able to remember all the verses, they made up their own with amusing results. They amused each other with suggestions of Kipling's company on this trip and agreed they might write to him at a later date. It was a mental exercise lifting them for a while out of the desert.

Little hills grew scarcer but they wound from one to another ever alert for gold and minerals, water or footprints until at last they stood on a sandstone ridge and considered.

Three days march to the south-west, guarding their approach to the Western Australian border, stood the mountains Leisler and Strickland. In 1899, Tietkins had recorded rockholes there. Due west, barely visible, a rounded hill surfaced on the horizon its very shape suggesting granite, a more likely place for water considering the run-off during rain. Ben suggested that it was too big a gamble, perhaps fatal, to chance the distance to the mountains, and the camels were still not in the best condition. It would be wiser to turn back to Davenport, this time to Eva Springs, the nearest of Davenport's many waters.

Reeds and rushes encircle these springs where clean water dribbles out of a gorge in the hillside. Because the camels' feet were sore and blistered, they were attended to and given three days' recuperation.

Aboriginals approached displaying none of the usual caution and diffidence. Instead, between gales of uninhibited mirth, they suggested that the travellers had got themselves in a right old mess. It was obvious to them by the many tracks going in different directions that these foolish pale people had got themselves into the desert and lost themselves trying to get out.

Amused, refreshed and a little chastened, the team left the luxury of Eva Springs behind them and headed due south with the aim of finding Thomas Reservoir. They crossed their earlier tracks near Mount Stanley and made excellent progress through occasional bands of fresh green grass and herbage. It was clear that there had been scuds of rain, but none sufficient to leave any water in the rockholes en route. There were, however, sand-hills and a towering one so steep the camels were severely tested. Challenged, 'Darkie', the obvious team leader, displayed what could have only been genetic memory. He dropped to his knees, keeping his hind legs straight and carrying his packs at the level, shuffled up the steep incline. None of his team-mates followed suit and made very heavy weather of the climb. A sandstorm of epic proportions flayed them. They muffled hands and faces and were forced to sometimes halt and hoosh the camels facing them away from the driving force. Bursts of hail mingled with dust and gravel and, were they fanciful, they could well have imagined themselves not welcome here.

There wasn't much water at Salvation Rockhole, by now an old friend, but sufficient to hold them together for the 34-mile hike south-south-westerly to the Cleland Range.

One Stormy Night

Locating Thomas Reservoir in these hills they luxuriated in a clean, deep pool surrounded by wild fig trees and overlooked by a gallery of native art. Every water supply successfully discovered was a triumph of such life-giving proportions that they were able to feel momentarily invincible. Any major water such as this and among such picturesque surroundings was seductive once they had attended to their chores and recuperated. An aspect they were reluctant to leave was the cool shade. Nevertheless they headed out and turned west on the third day and searched without success for Worman Rocks, discovered by Teitkins thirty four years previously. Their compass readings obviously didn't correlate and neither did his description. Finding no signs of the springs described by the explorer, they had no alternative but to once again retrace their steps.

They ran into a veritable forest of quandongs and both men and beasts ate their fill. They then crammed every box and bag with the fruit to carry away with them. Still they rode through threatening thunderstorms and still it didn't rain. They were led again by natives to an undrinkable waterhole and with their recent near-perish still very much on their minds, their border disappointment was not a priority.

From the vantage point of a higher sandhill, looking south they could see distantly an arm of Lake Amadeus (now known as Lakes Neal and Hopkins,) and further, Mount Harris, the highest peak in Blood Range. They set their sights in that direction.

Still the weather oppressed them. As they neared Mount Harris the storm clouds burst asunder pouring forth the most violent storm anyone could remember. They quickly unloaded and covered their possessions, secured the camels and huddled beneath a square of canvas through which torrents of water ran. When night closed in, they were still immobilised, spellbound by dramatic lightning and almost deafened by thunder. Finally in a crashing crescendo a great flourish of lightning pierced the very point of Mount Harris and a shower of blue sparks erupted followed by a glow which hung suspended above the peak. The storm hushed as suddenly as it had begun and slunk away into the night as if afraid of its own power.

Now there was a whole new luxurious world, awash and almost overnight turning green. Mosquitoes and snakes. As the soil dried, a moving carpet of black ants harassed them and neither man nor beast dared stand still. Mount Skene, a twin-peaked prominence of the Petermann Ranges, beckoned. They marched purposely feeling rejuvenated, another perish averted. In Ben's words, they were 'uncomfortably invincible'.

Among the spinifex, snakes increased to almost nightmare proportions and they watched helplessly while King, their most stoic camel, died within brief minutes of being bitten. Michael, studying the bite, carelessly touched some seeping poison which ran into the cracks of one of his fingers and made him seriously ill. Ben walked beside him, leading his camel and encouraging him. To take Michael's mind off his predicament, he recited his favourite recently written poem by Dorothea Mackellar. 'I love a sunburnt country....' As Michael's symptoms eased he asked for repeats until he too knew the verses by heart.

Finally arrived at Docker Creek, carved between the Petermanns and Blood Range, they spent three days on a waterhole and took stock while the patient recuperated.

5
Third Time Lucky
1932

Another Attempt to Cross the Border

A check of their remaining stores and hopes of more plentiful wildlife since the rain gave them the confidence to make another attempt to cross the border towards Laverton. There was every reason to believe that this time, travelling with the mountains, they would be better favoured. There was also the old adage, 'third time lucky' to boost their confidence. They wound around the end of the Petermann Ranges and crossed the Western Australian border on October the 23rd travelling parallel with the Schwerin Mural Crescent. Princess Alexandra Parrots escorted them and enchanted them. Knowing these birds were unscrupulously trapped and sold for large sums, it was heartening to find them in such large numbers out of harm's way.

At the foot of Gill's Pinnacle they located Gordon Springs. A little waterfall splashed playfully down from one overflowing rock-basin to another. Eucalypts offered generous shade and security to hatchling birds and native figs dripped fruit. It was another difficult place to leave, but leave they must. Although provisioned for a long trip, they had to ration themselves and supplement with bush tucker and game when and where they could. To have loitered unnecessarily anywhere would have been foolish.

Buried amid the kaleidoscopic background of this trip was a knowledge which surfaced now. Frequent footprints alerted them which Ben read and concluded that two raiding parties were stalking each other and it had to be the Petermanns and the Rawlinsons. In the previous year two prospectors had disappeared near Sladen Waters. When their camels turned up many months later near Alice Springs, without packs or saddles it was suspected they had been murdered. The evidence pointed to the Rawlinson tribe under the leadership of one elder known as Tiger.

An Awkward Situation in the Rawlinson Ranges

Now approaching Sladen Waters, in the middle of the Rawlinson Range, the cameleers halted and camped within a circle of baggage with rifles at the ready and took turns watching but it is doubtful whether anyone slept. Lockie and Jack dragged their swags a little apart, as was their habit, but within an hour after nightfall, their bloodcurdling screams brought the other three to their feet with rifles at the ready.

It seemed that two kadaitja-booted warriors had crept towards the camp but had failed to notice Lockie and Jack until they tripped over their swags. All four erupted in vocal terror and the would-be-attackers fled.

If Ben's reading of the situation was correct, they were caught in an awkward position. Between the enmity of two warring parties the camel-men presented an easy target. Because neither of the protagonists made any attempt to kill the camels, an easy mark, it had to be assumed they were frightened of them.

They moved cautiously.

Mosquitoes drove the camels mad and almost blinded them. They bit into the camels' eyelids, which swelled out of all proportion presenting a larger area for further assault. Their eyes ran continually and attracted persistent swarms of flies. Aggravated by dust they became hot and painful. Nose to tail they continually sneezed and shook their heads trying to rub away the offence against their neighbour's flanks. As their eyesight grew worse they stumbled and blundered, injuring foot-pads. They threw tantrums and refused to stand when loaded. Castor oil massaged around the eyes gave them their only relief. They leaned their heads into Ben's soothing hands and thanked him with appreciative grunts, nudging his shoulders for more attention when he walked away.

A searing heat oppressed them as the team again weighed their problems. They chose to trek by night hoping to lessen any danger of attack.

The moon was thin; once set it was difficult to see. They would occasionally throw matches into a spinifex patch to light their way, until eventually by daylight, they saw frequent rising smokes indicating that they had passed through the danger zone. This was Lasseter country where his reef was supposed to have been. They kept an eye out for signs of auriferous quartz.

Raiding parties would never have advertised their own positions with smoke signals so they were able to relax their guard somewhat and were travelling more comfortably when they arrived at Circus Waters. A major thunderstorm preceded them and waterfalls and further flights of parrots enhanced the time they spent there.

Circus Waters and Sladen Waters are intermittent springs which flow near the end of the Rawlinson Range. They were the starting point for Giles' ill-fated trip in 1873 into the nearby Gibson Desert where his companion Gibson is thought to have perished although no body has ever been found.

In his box Michael carried a copy of a diary of the earlier Hill's Border Expedition, which had ended in tragedy when Norton, one of their party, was fatally speared. Subsequently an Afghan camel-handler went berserk and had to be shot in self-defence. The survivors buried all but the bare essentials before abandoning the quest. That had been thirty years prior but Ben was able to locate the cache, and found it had been raided and destroyed probably almost immediately after interment.

A Message from Lasseter

On the 2nd of November they left Circus Waters and trekked towards the Alfred and Marie Range, travelling well through good feed and herbage. Here too grew a plethora of Emu plums which were a welcome addition to their diet and great flocks of Princess Alexandra parrots objected raucously to sharing them.

At Lake Christopher they were astonished to locate on a tree, among cabbage gums and desert oaks, a sign which read, LASSETER, 2-12-30. Towards Tommy's Flat and within the perimeter of the dry mud lake they came upon a message etched with a stick, reading DIG UNDER followed by a further indecipherable word.

Nearby was evidence of a stake having been left propped up by three forked sticks and fluttering remnants of cloth. Careful to leave the message undisturbed, they dug and searched to no avail but located remains of a camp which produced nothing of any interest except a few rags.

They deduced that here Lasseter was two hundred miles away from Piltardie where he claimed in his journal to have been on that date.

The camels seemed to have agreed among themselves to make a break homewards and it was difficult to keep them in check. When in camp, they had to be hobbled and sidelined so they could feed but were unable to cover any distance. They sulked.

Summer's onslaught was debilitating and the camels refused to budge when they found a shady tree; they broke nose-lines in rebellion and voiced their opinions both ends.

The party were now out of the mountains. Their trials and tribulations had been assuaged by the astounding scenery through which they'd passed. Giles, and perhaps Lasseter, had preceded them in this isolated part of the world and they almost felt accompanied by something of those men.

111 Miles to Water

Having searched for Tommy's Flat, 'rockholes on a grassy mulga flat' described by Ernest Giles, they were forced against time to give up. 'The latitude of it is 24 degrees 52'3. It bears 9 degrees south of west from a peculiar red sandhill that is visible from any of the hills at the western extremity of the Rawlinson Range; and lies in a flat or hollow between the said red sandhill and the nearest of a few low stony hills about four miles further away to the west.' Their failure put them at a disadvantage as there were 65 miles to go to Springs Granite, (Granite Springs), with only three days' water supply left. The heat was a more serious problem than it had been in their retreat from Lake MacKay and they were not as fit as they had been then. Their only driving force was will-power. They would rest in shade in the middle of the day and again they were forced to travel by night setting their course by the stars.

When they wearied of the burden of this second desperate search for water, Michael discovered enough in a gnamma hole on a low break-a-way ridge to liberate the camels from the worst pangs of their thirst.

Stan was about sixty years old and the two younger men were continually amazed at his stamina. He was no stranger to the bush, having been a government assayer for some years at Arltunga. He had withstood without a stumble all the rigours of the trip. All three of them had become whip-thin, and had pulled in their belts to new notches. Their camels' humps were sadly depleted and their urinary output, without sufficient water, had become highly acidic. Where it trickled down their legs they broke out in ulcers and the weary, worried men had extra duties attending to their sores and wrapping them against marauding flies. There was not enough rest for either men or camels but they couldn't let up. It was with an almost superhuman effort that they managed to keep moving.

Elder Creek appeared and beneath its leafy trees they stumbled upon an old well. The water was fouled with the rotting corpses of many dead birds and animals but such was their need, they were easily able to ignore that inconvenience and found renewed strength to bucket for the many hours it required to water the grateful camels. They had come, in this last dash, 111 miles in three days and four hard-pushed nights with little more than twenty four hours' sleep.

Warburton Junction Waterhole

Three miles further up the creek at the junction of the Warburton, they found a more wholesome water supply and settled down to recover and attend to the camels. It gave them too a much needed opportunity to bathe away the sweat-encrusted sore points on their own bodies and wash their clothes. Footprints, smokes and calls presented evidence of a large Aboriginal gathering and when Ben enticed a small group to communicate, he discovered it was for a three-fold purpose. The men were engaged in hunting down and punishing a group of wrongdoers. That done, they had an initiation ceremony to make young men and would eventually gather with their women-folk for a great corroboree. It was a busy sphere of activity and that such a large body of men made so little sound would have been unnerving were they unaware of the purpose.

Other information, hesitantly relayed, was that in the previous year, as nearly as could be ascertained by new-moon count, a 'red nose' (a white man) had been camped nearby. Attempting to show his friendly intent, he had offered meat which was strange to them. They had taken it away to study it more closely and concluded it to be the flesh of one of their women who had gone missing. Ben's assumption was that it may well have been bacon which most bushmen carried. The Aborigines refused to take him to where the body was hidden and the subject was closed when he sought further evidence, Ben, Michael and Stan were in too isolated a position among so many to push the point.

They broke camp and changed their direction towards Paura because the same Aborigines had insisted that Granite Springs was dry. It may or may not have been, but it was obviously out of bounds to them in this current situation.

They were much recovered which was as well because the heavy rainfalls had not extended this far and what there might have been had withered in summer's intensity. Indeed, so dry was the country that they were amazed to see rabbits climbing trees for want of any foraging at ground level.

At Paura the team took the opportunity to make a more comfortable camp while they gathered themselves together for the final push to Laverton nearly 400 miles away. They were all busy. There were journals to catch up with, ropes to splice, pack and riding saddles to repair, camels to treat and checks to make on their remaining gear.

Unexpected Company

They, too, needed the rest and recuperation. They knew how fortunate they had been to have scraped through without any serious ill-effects to this point and were intent on upgrading their equipment when someone imagined the sound of a motor.

When a low wind moans through mulga trees, the sound produced is very like a distant motor. The ears become almost painful, straining to pick up the sound again and identify it. When eventually there is no doubt it is definitely a motor, then the ears are still channelled to the exclusion of all else. What the devil would a motor be doing way out here?

Stewart and Cable pulled up in a cloud of dust, much surprised at the sight of a camel-team in this distant outback. They were the ground support vehicle for Keith Farmer and Errol Coote who would fly out prospecting, in a Sopwith Gnu from Laverton. For them a mere four-hour trip. It was getting crowded but it was a welcome diversion.

They sat late into that night and the newcomers offered fresh eggs and bacon which, lubricated with beer, somewhat surprised their tastebuds. There hadn't been any other than local game, bully-beef, damper and tinned fruit for quite awhile. It was good to have news of a great wide world from which they had been shut away far too long, and with their rations running low, the meal reminded them of more sophisticated fare to be had in Laverton.

They farewelled their visitors and with renewed energy set off the next morning. The heat drove them back to night travelling. It had become a habit and they had overcome the associated problems. But the camels didn't always agree and were still unhappy about the direction.

While the men reminded each other of hot baths and fresh bread slathered with butter, of clean clothes and soft shoes; the camels, in indignation, broke wind and raged vocally. Michael's mount grew more and more frustrated and difficult to handle and finally went berserk. He raged and flung himself into high contortions, threw himself around trees in an effort to dislodge his rider, bellowed and screamed his rage, but Michael hung on.

He never quietened until he had exhausted himself, by which time the entire string of camels were confused and anxious.

Ben, on foot, ran to catch and soothe him and to congratulate Michael on the fanciest ride he had ever seen. Michael was, for a while, too shaken to dismount and when he did, it was to discover that in the struggle he had twisted his back. The camels were rested and comforted while the billy boiled, then, with help, Michael climbed back into the saddle but Ben led by the nose-line until assured there would be no repeat performance.

Michael is Injured

Only a couple of days later a snake slid out of the undergrowth. Michael was walking to give his back some ease and mindful of the dreadful death of 'King' from snake bite, without thought, threw himself towards a heavy stick. He swung it mightily and destroyed the snake within inches of his camel. But unbelievable pain stabbed through him and down his spine and he knew he was in dire trouble.

His mates managed to lift him into the saddle but it proved too agonising. Carefully and with great concern they eased him back to the ground and made him as comfortable as they could.

They were fortunate in that they knew somewhere behind them were their companions of the survey relief truck and they were due soon to head back along the same route to Laverton.

They could only make camp and wait.

There was a disquiet in their waiting. It was possible that the truck might take another route back. It had been by pure chance that they had crossed each other's paths but it must now be expected the truck would look for them on their return. Ben was naturally of a positive nature but now while Michael's situation was desperate, he was inclined to think of other outback instances. Very recently, a toddler, born with a genetic problem, had fallen into an open camp-fire and badly burned his face. Expecting that blindness would result, the father decreed the poor mite had enough against him without such a double handicap and should be 'done away with'. So far from medical help, in the middle of a national depression, it was an agonising time for the mother who used all her nursing skills and feared to sleep. Fortunately the child's burns healed well and his eyesight seemed unaffected.

Perhaps an arrogance, maybe a blind faith, had to be an ingredient in the make-up of pioneers.

When the motor did arrive, relief was so palpable that only then were Stan and Ben aware of just how anxious they had been for Michael. With every item of comfort donated from every swag, they made up a bed in the back of the truck for the patient and lifted him carefully on board, waving him off to qualified care.

Watching the truck until the last signs of dust had disappeared they broke camp and set off the last two hundred miles, re-uniting without incident on Christmas Eve at Laverton Hospital.

A Bushman's Holiday

> '...Ben was tough. He was a top-notch bushman. He was young but he had more bushcraft than most old timers. He was Centralian born and bred. Resourceful, work-hardened, weather-beaten, astute and careful when need be, boisterous and downright reckless on a spree, he was the epitome of the pioneering Australian Inlander.'
>
> Michael Terry. September 8th, 1965. *People Magazine.*
> 'Portrait of a real bushman.'

They found Michael had sustained such serious damage that he had later to be removed to the Royal Adelaide Hospital. While still in the Laverton Hospital he was visited at a later date by a somewhat physically damaged but still cheerful Ben. On the previous evening Ben had found convivial company in one of the local hotels and happily agreed to help make up the numbers for a challenge against the football team from the neighbouring township, Morgan. Both teams celebrated afterwards in the same hotel and at closing time Ben insisted loudly, 'You Morgan boys can't play football and you can't fight, either.' Neither side won the ensuing battle but they flayed each other mercilessly and enjoyed the confrontation.

Visiting Kalgoorlie, strolling one morning along Hannon Street, Ben came upon a small lad crying his heart out. Ben hunched down and sat in the gutter with the boy. 'What's up, mate?' he queried. It seemed the boy's unregistered puppy had been arrested by the town dog-catcher. Seeing a man he knew riding a motor-bike towards them, and having filled him in on the sad story, the two located the van and hauled out the catcher. While Ben kept him in a neck-hold, his friend opened the back door and released the day's dog-haul.

Later in court, the magistrate, laughing, dismissed the charges.

Ben bought the friend's motor bike intending to ride it all the way back to Alice Springs, a two-thousand mile trip. Tiring of the really rugged road, he took the bike up onto the railway track and travelled well until very near to a little rail stop he hit a cattle-pit between the rails. The bike fell into the pit and Ben sailed over it. Leaving the wrecked machine he caught the train to Port Augusta where he transferred to the Ghan. One of his friends was on board, celebrating his birthday en route and by the time the train arrived at Oodnadatta it was a pretty willing party. Ben, a little the worse for wear, took the opportunity while the train shunted and loaded to step off and take a stroll to get some fresh air. Looking up later, he noticed the red lights of the departing train and knowing there would be no other along for two weeks, took off after it. Managing to haul himself onto a buffer he rode it all the way to Alberga, the next stop.

My Mate, Dutchie

Hatches Creek, the mining area in the Davenport Range north of Glen Maggie, was a magnet to Ben. He spent his spare time there catching up with mates and prospecting. On one of his trips he met a large young man, of very much his own build and age and felt some empathy, especially when he discovered the young man was broke, hungry, looking for work and was hitch-hiking. He had jumped ship in Darwin. Ben took him back to Glen Maggie and Dutchie became part of the family. His English was very limited but with the aid of a dictionary which he kept by him at all times, he was quickly able to make himself understood and eventually spoke in his rich baritone without any accent at all.

He was well liked and settled easily into the Australian outback without any obvious backward glance at his family and his home in Holland. But the children of the Centre knew he missed another place for he was a master story teller and in his tales he gave himself away. He was the little boy who put his finger in the dyke and saved Holland. His fairies and pixies were Dutch. It was a Dutch landscape of tulips, windmills and sabots he coloured and set his stories within. It was a Dutch Troll who lived in the cupboard beneath the washing-up dish and who would punish horribly unless the dishes were perfectly washed and put away.

He followed Ben, or led him, or perhaps it is fair to say, together, they enjoyed many impulsive indulgences. They shared a sense of the ridiculous and a thirst for adventure but Dutchie could never be prevailed upon to accompany Ben on any one of his bush trips.

They also shared, with so many others, the quest for gold. The tent-towns and mining camps throughout the inland drew them like magnets. They were undisciplined settlements and none more so than Tennant's Creek. When an Aboriginal named Frank handed a pretty piece of stone to the operator in charge of the Telegraph Station near the creek, he in turn showed it to two prospectors, Jack Noble and Ralf (or Ralph) Hadlock. That piece of stone triggered the gold rush of 1932. Joe Kilgariff in Alice Springs loaded up his truck with rations and grog and the bare essentials to build a store near the Telegraph Station but it rained and he got very badly bogged. Hungry and thirsty men besieged him, so sensibly Joe erected his store beside the bog, twelve miles south of the Telegraph Station. The township grew up around his store.

It was a 'shoot-em-up' town. When the first town blocks were thrown open for public sale, the acting sheriff fired a gun and would-be buyers leapt to peg their sites. Rifles and hand-guns were carried or worn to protect personal property. Battles were fought over two bottles of whisky, or mining claims or even over ownership of firearms.

Ben was never known to carry or wear a gun in town but on one of his ventures to 'The Tennant' he found himself caught up in an affray which ended in murder. He and Jack Noble were acquitted of involvement after a dramatic four-day trial in Alice Springs. In February, 1935, it was the first sitting of the Supreme Court ever held in the Alice. A third man confessed, 'Yes, I shot O'Brien in self-defence.' The jury's verdict of 'Not Guilty' was unanimous.

6
A Second Expedition with Michael Terry
1933

A Bogged Camel

On April the 21st 1933 a recovered Michael with Stan met Ben at Brookes Soak. Ben arrived with two Aboriginal offsiders called Bob and Paddy and a string of nine camels. Michael had stores and equipment ready for this new trip and because there had been wide-spread rains twice in the interim, this time they expected none of their previous problems.

Michael was the boss-man, Stan the Geologist and Ben was guide and camel-man but since their previous trip they had settled into an amiable partnership. Apart from natural history details they were three men seeking an Eldorado and much of the country was already familiar to them.

It was an easy 80 miles to Mount Doreen, a comfortable bough-shed homestead where Bill Braitling with his young wife and infant son was labouring to establish a cattle station. While they were there a great black sky lowered and poured inches of rain down upon a landscape already saturated and over which they planned to travel.

It was awhile before they could leave and perhaps they were a little too eager at that because, before they had covered many miles, one of their camels went down in a bog, and went deeper by the minute, sucked down by its own desperate struggles.

While they could still reach the packs, they unloaded and quickly chopped logs and used them as leverage to halt the descent until further logs could be chopped and brought into the battle. The quicker they worked, the more the camel struggled and sank, but they laboured on and finally dragged free a giant, agitated mud-ball. Their last trip had been a struggle against aridity; was this one going to the other extreme?

Their direction was set for Vaughan Springs and when Bob and Paddy realised they would pause at Pikili, they panicked. It was evident that when they had previously been there and noted the absence of Aborigines, they had guessed correctly that it was a taboo place.

Meanwhile, having effortlessly arrived at Treuer Range, they trudged to its north/western end to Comatchee water hole and were entranced with the prettiness of its setting but having drunk, they pushed on the extra miles to Vaughan Springs. A brighter jewel in the chain of memories of their previous trip, it was as rewarding now as it had been then. They found Stan's date seeds had turned into a copse of about three dozen infant palms and while the others set about the usual camel and camp chores, Stan turned gardener and relocated those plants he considered too closely grown. It must truly have been an amazingly rich plum duff he'd produced at that other time.

In a westerly direction from Mount Davenport, they spent two weeks in an intense prospecting activity and while camped on Midginbanda Waterhole, discovered such a plethora of emu eggs that they were happily able to indulge in scrambled eggs and omelettes. With such a welcome and nourishing addition to their diet, their expectations of this second venture seemed thus far vindicated and they had the time to indulge more freely in their main purpose, prospecting.

All three men had their own specific duties but they were each capable of and frequently did assist each other and consequently they were ever alert to every aspect. Only when night fell, did they relax, light up their pipes and chat of other things.

Return to O'Grady's Well

Towards Lake MacKay, they came again to Blackshaft and found it full thanks to the effort they had put in on first acquaintance. Then they had dug deeply and without result but now it brimmed.

On to Mount Surprise and another wonderful supply of water and still the camels could pick and choose what they ate.

They erected a pole and inscribed their name for the place together with the date and their initials before pushing on to the McEwin Hills. The camels were so well watered and fed that they became frisky and overly energetic. It required every effort to keep them in check. They jumped at their own shadows and skittered off at tangents breaking nose-lines and confusing their mates. They dragged their loads against tree trunks and scattered contents. All Ben's skills and everyone's patience were overworked by the time they reached O'Grady's Well.

Where they had previously laboured in a desperate 35 foot delve for a pitiful supply of filthy water, they now found it awash. Still the camels pranced but they had a job ahead of them and hard work would surely settle them down.

Back to Lake MacKay. They tried to cross the salt crusted surface but the canteen camel broke through into thick slime. It was a long and difficult exercise to get it back onto dry land and they had to retrace their steps and take a more northerly arc. On the 31st of May 1933 they crossed the Western Australian border and finally arrived at the mountain range they had been forced previously to turn away from.

It wasn't marked on any map which they found most odd because it was a very noticeable range. On their earlier sightings they had discussed it and agreed that should they achieve its heights they would name it after Alec Ross, the only still-living member of Ernest Giles' expedition.

They left a cairn and enclosed in it a tin with relevant information. From this high vantage point, the country spread beyond vision and imagination and the men wondered that they, in their insignificance, had dared to venture into it and that they had so far survived it. Enclosed within each day's tasks, it was here, for the first time, they became aware of the vastness of their enterprise and felt chastened.

Bit by bit, day by day, they notched up the plodding miles and noted them down in their journals. There had been no Aboriginal signs but now they became aware of surrounding smoke talk, sent spiralling upwards and quickly ended. Their presence and their route were being watched and monitored. Bill Braitling had warned them that the Boonarra were a people who could be dangerous and now they had crossed into their tribal country.

They would have to depend on Ben's knowledge of protocol and taboo and his communication skills. Three warriors appeared against the skyline and by gestures indicated a willingness to talk. The three men walked out to a half-way point flourishing empty hands, before squatting down in relaxed positions to await their visitors. The warriors approached and laid down their spears and woomeras. A long silence followed while everyone practised friendly facial expressions and waited for an opening.

Finally, everyone spoke at once and the tension was broken. Their's was a different language from any Ben knew but he was able to tune in to it and with a combination of finger talk they were able to communicate. Ben had an extraordinary facility with languages of not only Aboriginal but also foreign derivation and communication was therefore never a major problem for him. His world had always been one of fewer English speaking friends than otherwise and his ability had come to him naturally. His environment was filled with such interesting people that he couldn't conceive of such a thing as language barriers.

Now he discovered all that the natives were willing to impart of feed and waterholes and was able to assure them that this small collection of men and camels, (emu-horses), were travelling peacefully with no ill-intent. Small gifts were exchanged, hand-mirrors packed for such an event, in return for charred goanna and berries.

Boonara Country: The Debil-Debil and the Beserk Camel

Pushing on through a verdant landscape, the team crossed and examined camel tracks which Ben judged to be seven or eight years old and probably left by Jimmy Wickham prospecting out in this neighbourhood with a team of Government camels. He had returned with a large lump of gold and had been unable to rediscover his source. Regrettably, he had been incarcerated and in that period of time his journals were stolen by someone who seemed to believe himself more deserving. The reef was never relocated.

In conversation with Jimmy, Ben had learned of a large freshwater lake, but found its existence difficult to believe. Now nearing that location he was alert to the possibilities.

While fossicking around the western slopes of the Alec Ross, an Aboriginal trio approached and directed them to a rockhole known as Kooala-Nynma.

It was particularly sacred and they could water there but must be careful.

Wishing they had known about it the previous year, they were told it had then been dry.

Carefully and almost piously silent, they topped up their canteens and led their animals in to drink their fill before retiring a decent distance to camp.

After five days of hard prospecting, they crept away, still under close scrutiny. This was the debil-debil country Bob and Paddy had panicked about and their fears were mounting until they became a serious problem. Their one job was to follow the camel bells each morning and return the team to camp and assist with the loading. Now, however, they wouldn't move beyond arm's reach of any one of the white men and Ben had to walk out with them, carrying his revolver, so that they felt protected.

They dogged his footsteps and jumped at shadows. They bumped into each other and into Ben and when he laughed, his sudden roar of mirth unmanned them totally and they fled back to camp.

The locals had embroidered their initial fears with tales of a monster who tore out the stomachs of the unwary and ate those parts while the victim watched. His name was too terrible to mention and if spoken would invoke his presence.

Within days the camp woke to a world obliterated by thick, white fog and Bob and Paddy were certain it was the doing of their demon. It was indeed a strange phenomenon in this location and particularly under current weather conditions. It lay heavily and soaked into everything, food, clothing, swags, and boots and dripped wearily from branches. When it lifted, finally, in mid morning, they were too saturated to break camp and were forced to dry everything out before attempting to move on again next day.

Heading north west, they came to a vision of five large sheets of water lying in deep marsh country and hedged by ti-trees. Wild-life abounded.

So did lurking Aborigines but they were not invited to approach because Bob and Paddy were already too overloaded with horror to absorb any more.

Another one of the camels exploded for no apparent reason in a paroxysm of evil intent. He gyrated and twisted and splintered his pack-boxes, rushed against trees and smashed off his load. That done, trailing ropes, he headed away into the scrub bellowing abuse. Meanwhile his companions ran about, breaking fresh nose-lines, bumping into each other and joining in the cacophony. Retrieved, it was a sore and sorry camel. Ben chopped down a tree-limb and carved new pieces for the broken pack-saddle while everyone else collected scattered food supplies, mended boxes and generally made the best of the situation. The miscreant, tied up in isolation, sulked.

While Ben carved new nose-pegs and mended broken lines, Stan and Michael went about the business of making camp and getting a meal together. Ignored, the villain whimpered occasionally like a sorry child.

This was a young bunch of camels on their first trip away from home and they weren't much interested in earning this sort of a living.

Discovery of a Hidden Paradise

From a high sandstone ridge, the travellers could see another distant lake and a great depression sunk quite a lot lower than the rest of the landscape.

As they neared it seemed ringed by sandstone cliffs and ghost gums. Above flew white-winged corellas and finches, an obvious indication of water, probably permanent. Michael rode on ahead and was amazed at what was spread before him. He edged his riding camel down an incline into a valley wide and verdant, so thickly carpeted with grasses, herbage and shrubs that the ground was barely visible. Wattles and ti-tree cast deep shade and it was with difficulty that Michael pushed his camel through. Suddenly, at his feet was a waterhole, and beyond it another five. It was all so lush that he sat in amazement for some time before scrambling down from his saddle and casting around for wood with which to light a fire beckoning the others. There seemed to be very little dead or dry wood to be had and it took him awhile to get a smoke going.

While he waited, he stared at the impressive sight before him and thought of friends in Adelaide whom he much admired and how they would enjoy this.

Without a doubt, he would name it after them, Brookman Waters.

Here was an expanse of wonderfully pure, clear water, not guarded by Aborigines and when the others caught up with him they found Michael luxuriating naked and blissfully relaxed in the waterhole.

It took only seconds to join him.

When he set about surveying the area, Michael calculated that the depression which constituted the basin was roughly 800 feet below the average level surrounding it. It was protected from the usual drying winds which scoured the higher plains and from fiercer evaporation, and within its protection all green things flourished.

There had always been rumours of such a place, known among Aborigines as Chugga-Kurri but it had seemed mythical and to an extent, impossible.

Aborigines near Broome had spoken of it and pointed south east. Down across the Nullabor a vague directional wave indicated north. To stumble almost accidentally into it was beyond credibility.

The team prowled, measured, took samples and photographs and discovered a well-watered creek where grew reeds and Moreton Bay Ash.

The watercourse ran into the basin from a south-south-easterly direction and where the bigger trees stood back, thickets of wattle hugged its margins.

From a prominent lookout they viewed an amphitheatre to which they gave the name, Redcliff Pound and the creek winding through it into the basin, Nicker Creek. The entire forty-by-twenty mile salad bowl required a more seriously considered title, but after a great deal of discussion they settled on the obvious, Hidden Basin.

When they reluctantly headed out, it was towards a prominent range to the north east. Looking back, they were not for long able to see any sign of their valley and had to agree they had named it well after all.

Four Days without Feed

Salt lakes, water holes, more hills and more hills, a running stream of fresh water winding into the saltiness of Lake White. Wet bog country reminded them all of artesian landscapes and they wondered about mound springs but were unable to locate any. When their canteens lightened, Michael discovered a pool of fresh water edged with something resembling a four leafed clover and named it Lucky Pool. For four days they camped and scouted, adjusted and redistributed pack-saddle weights and mended.

Michael and Ben climbed a peak nearby to get their bearings and found chipped in sandstone, the initials, J.R.O.D. and the year, 1910. They thought they knew all of the few crossings made in any of this western desert country but these were strange initials and they could only assume that it had to have been someone to or from the Tanami gold rush. They hoped he had fared well. They left their own initials and the date within a cairn as much as a memorial to the unknown traveller as to perhaps keep his spirit company.

They were aware that an Afghan had wandered in this country on his own, having been involved in a dispute with his countrymen in the vicinity of Broome and had succeeded in crossing to Alice Springs where his reports of Lake MacKay predated its discovery by Warburton. But his would not have been these initials. It was a mystery, still unsolved. Today one wonders if perhaps they were the initials of one Jim O'Donnell, prospector of Western Australia.

Since leaving Hidden Basin the team had outwalked rain-enriched country and found themselves once more in the lap of desperation. Camel-feed grew less available until it ran out altogether and their only water had been the discovery of a soakage at the base of what they called False Mount Russell. Camp by camp the camels were forced to walk further for meagre pickings until, running out of patience, they tried heading back in the general direction of Alice Springs. Four days without feed and Ben knew they couldn't go another. There was only enough water in their canteens for the men, if taken sparingly.

On the fifth day they found a few bloodwood trees growing along a dried-out water-course but the leaves were enough for the camels to feed on and would have to suffice. It was fifty miles to the Granites goldfield and Ben found himself in familiar surroundings from his desert crossing with Joe Brown, ten years before. He knew there would be an old soakage a little more than half-way and his memory was rewarded. The soak was there but its supply was barely adequate. They were forced to walk through the night and caught up with Michael who had preceded them to the Granites.

Simon Rieff and Black Cow

Michael was camped with Simon Rieff, an ex-Cossack who had been forced to flee the Russian Revolution. He had arrived on Australia's eastern shore by way of China knowing nothing of the English language and owning no more than a change of clothes. Working his way across Queensland, he arrived at Tennant Creek when gold discoveries there opened up mining jobs. He liked the people and the place and invested his wages in prospecting expeditions throughout the Centre. His Cossack background had inured him to hardships and outback conditions didn't bother him. He crossed paths with Ben and a mutual respect grew out of their shared interests. While Ben couldn't hear enough of Russia and Simon's experiences, Simon sought to learn all he could of the bush from Ben.

Simon's interest in geology took him far afield. He accompanied Doctor Madigan into the Simpson Desert and led Madigan and Douglas Mawson to a nitrate deposit he had discovered in the MacDonnell Ranges. He accompanied Madigan on most of his inland trips and developed much expertise. He then settled comfortably into marriage and fatherhood and adopted the inland as his home. Simon enjoyed Michael's company because he too knew something of Russia. It was a welcome diversion for both men.

Having garnered outside news, sent off telegrams and refreshed themselves with company other than their own, the team were in a hurry now to reach Alice Springs.

In the general direction of home and in more relaxed mode, they weren't expecting trouble. 'Black Cow', the camel carrying their explosives, was up until now most reasonable, but without warning threw a tantrum. As she bucked and screamed in frenzied attempts to rid her load, the men froze in horror. Ben, taking his life in his hands ran to grab her lines. She evaded him and threw herself among her companions kicking viciously at her loosening ropes and flailing baggage until it all surrendered to gravity and smacked to the ground. Nothing exploded.

Ben grabbed a mount and with Michael, tried to head her back the way they had come, but she was of one mind and that was to dump her load and go home. Her ropes flew behind her as she galloped, neck extended, southwards and although they raced along keeping pace and hoping she would tire, her determination powered her. When it grew too dark to see, they turned back with the confidence that Stan would have a good fire burning as a beacon.

The remnants of the load were divided next morning among the other camels and Ben, with Bob and Paddy, took the string to water at a near-by water-hole while Stan and Michael knuckled down to three days of serious prospecting but again without result. Reunited, the party returned by way of Wickham's Well and Whittington Soak at Mount Singleton, to Braitling's camp and thence towards Alice Springs. Their last lap of the journey was in a freezing, steady pall of rain and it was a wet and bedraggled group who rode into town at journeys' end on August 13th.

In this second, shorter trip over four and a half months they had covered, at a rough estimate, 1161 miles.

A Pair of Prehistoric Perenties and Ben's Bulldog

Ben never seems to have unpacked and put his feet up for long before he was off again on another one of his trips. There always seemed to be an explorer, a survey team, a prospector or an adventurer seeking his services. There is no record of his ever having taken a holiday but his family considered he was away so much 'little bit long way' that the times he spent with them were his 'breathers'.

He would stay awhile in Alice with his mother, enjoying the social atmosphere and catching up with mates. She was of course, always pleased to see him, but annoyed with some of his habits. Having warned him once to please come home quietly that night, and not to wake her, she finally heard the gate creek slowly open. A long silence followed while he obviously took off his boots and crept to the door. Another creek of hinges while the door opened and again silence. Settling back to sleep, Liz was startled fully awake by a loud double thump as Ben's boots hit the floor. He'd been too long in the bush.

Relieved of his washing and mending, he would head out to Undoolya and wander on to Love's Creek. He played tennis, caught up with family news and would suddenly be gone as unexpectedly as he had arrived, leaving boots and bits of clothing in a trail behind him. He would spend a few days with Gene at Todd River or wander north to see Claude.

There never seemed to be enough spare time for him to stay long in any one place. His visits were a breath of fresh air, a change in the routine and a time of bustle and amusement. He seemed to need his injection of family and friends to carry with him on his trips. For a while he had an English bulldog which he had discovered nearly perished along the main road. It had been obviously lost by a traveller and he was never able to find its real owner. It was a most unusual breed to find in the bush and its stance and face were so pugnacious that nobody wanted it or could get anywhere near it, but he and Ben suited each other. 'Dog' trotted always three paces behind Ben, never nearer because he had his dignity to consider. He hated camels and cars and cats and refused to be anything other than independent, as long as Ben remained in view.

Ben took prickles out of his feet and fed and watered him and had long conversations with him, otherwise it remained a distant affection. Ben's nieces and nephews were terrified of 'Dog' so he had to be tied to a long rope while visiting, usually half way between the homestead and the distant outside toilet. This caused a problem when the need arose so the trip would be put off until the last possible moment. 'Dog' would trot to the end of his tether and sit glaring while the child edged around in a wide circle trying to reach the goal before time ran out.

Most dogs living in the bush like to hunt but 'Dog' was too squarely built and 'too low in the wheel base.' He preferred to give the impression it was all beneath his dignity.

Ben took 'Dog' with him when they visited Gene at Todd River Station and found Gene not at home but camped on the southern side of Weeitcha Gorge with a few cattle. Wandering out in the evening light to bring in more wood for the camp-fire, Ben was confronted with a pair of giant perenties. He had heard about them but was inclined to disbelieve their size. He stood stock-still and his blood froze. When a perentie is frightened it will run up the nearest tree or person, clawing as it goes but this pair were a lot bigger than Ben was. Their size and appearance made him feel himself in the presence of something prehistoric. Their skin seemed black and scaly as if burned and they strolled arrogantly with heads lifted and tongues flicking, testing the evening air. Ben hardly breathed.

Suddenly 'Dog' spotted them and panicked. Forgetting his dignity, he fled between Ben's legs which wasn't safe enough so threw himself in a desperate heave towards Ben's chest obviously hoping for the safety of friendly arms.

Just when Ben had felt the danger passed, he now found himself in real trouble. All the unwonted activity alerted the perenties and they faced each other. 'Dog' trembled and tried to reposition himself on Ben's shoulders which upset the balance of the drama even more. While he clung to 'Dog' and pulled him away from his face, he was momentarily unable to see what the perenties were doing.

With 'Dog' whimpering and panting in his ear and nearly choking him, he was even unable to hear until a louder sound broke through. It was Gene. Knowing this was the haunt of these creatures, he realised there might be a problem and grabbing two large firesticks ran to Ben's aide. The perenties turned away unfazed but obviously not bothering to get involved with fire.

'But I've told you about those fellows,' said Gene.

'But who would believe it!' replied Ben.

'Don't you remember the bloke who nearly got killed when he had a battle with one on horse-back? The horse had to be shot.' Gene paused. 'That sure is one hell of a bush-dog you've got yourself there!' Gene could afford to be lofty about 'Dog' since his own dogs were staghounds and excellent hunters.

Actually, 'Dog' had been no fool to panic. At nearby Deep Well Station one such perentie had shredded a much larger dog belonging to Gerhardt Johannsen while the two were out hunting rabbits. Gerhardt had paused to sit down on an overhanging rock and survey the area. His dog sought the shade below the rock and a sudden battle erupted. Before Gerhardt could do anything, his dog was dead and he was himself face to face with the attacker. He fired his rifle and killed the perenti which surprised him because the suddenness of it all and the size of the perenti had made him tremble. He had the skin tanned because he doubted anyone would believe how big it was.

Mags Nearly Perishes in the Tanami Desert

In his travels Ben found tin on Coniston and turned the discovery over to the then owner, Randall Stafford. There is an equipped bore there now called Tin Field Bore, but no mine because the field was never quite a viable proposition.

These were still uncomfortable tribal regions and perhaps it was as much good luck as it was good management that they travelled with so little trouble. Ben was a rifleman of more than average excellence and the rifles they carried would have been in themselves a threatening protection. They would have been watched and bush telegraph would have marked their every move. The game they shot for food and their accuracy with the rifle would have been noted and commented upon.

There were still areas where Aboriginals had not yet come in contact with whites and it was wise to be cautious.

Mags married Rex (Kingsley) Hall and they had a son. With the two children they moved to Tanami where Rex worked on tribute on the mine. One hot Christmas when rations on the settlement had become too low for comfort, Rex and family set forth across the baking Tanami desert to buy supplies. They ran into trouble and this is how it was reported in the *Adelaide Advertiser* in December 1935.

Family's Terrible Plight.

Tennant Creek Goldfields, Thursday.

'Without food and water for four days, or clothing other than what they were wearing, Mr and Mrs Rex Hall and their two infant children had a terrifying ordeal in desert country on the road from Tanami to Halls Creek last week.

When their truck burst into flames near the Western Australian border 65 miles from Tanami last Sunday week, Mrs Hall threw her two children, the elder of whom is aged four and a half, among the bushes at the side of the track and jumped clear herself.

Nothing in the truck could be saved and all that was left for the family to do was to turn back on foot towards Tanami.

The Director of Mines in the Northern Territory, (Mr N.C.Bell) was at Tanami last night when Mr Hall struggled in with the news and said that his family had been unable to progress any further. He had pressed on so that another truck might be sent out to bring them in.

Mr Bell had gone on to Tanami from the Granites, whither he had been summoned by the Minister of the Interior (Mr Patterson) to inspect certain workings.

Near Collapse

Had it not been for Mrs Hall's amazing stamina, I doubt if she could have come through her ordeal at all, Mr Bell declared when he brought the news to Tennant Creeks Gold Fields today.

Mr Hall is working on tribute for the Tanami Goldmining Co., and set out with his wife and children for Gordon Downs in Western Australia, where he was to obtain meat supplies for the whole community. The truck became enveloped in flames in a flash when the exhaust became overheated plowing through the sandy stretches of road. Mrs Hall, a member of the Nicker family, famous throughout the Territory for two generations for their courage and pioneering in dangerous areas, helped her husband to carry the babies, but on Tuesday morning could go no further.

She was in a state of near collapse and the children were crying for food when her husband pushed on alone, reaching Tanami next night. A truck was sent out immediately to the rescue.....'

Aboriginal fire-sticks glimmered while they walked at night, parallel with them, and in daylight hours while they rested they watched wispy smoke signals spiralling among the spinifex. They had no idea what would happen to them when they became too weak to protect themselves.

Mags knew it was conkerberry season and where to dig for yulkas so with these the family were able to survive in a better manner than was suggested in the newspaper report. Conkerberries are thought to contain an high concentration of vitamin C.

When lacking shade among the low scrubby trees which grew there, she broke branches and threaded them crosswise thus casting a denser more comfortable shadow to rest beneath. She never slept, fearing that the children might wander off and be taken by the Aborigines who were none she knew. She didn't know this country like Ben did and wasn't herself the known entity he was. She continued, however, to hope they might appear and help but they remained beyond communication.

Mag's collapse was the result of blood poisoning due to the many spinifex needles which had pierced her uncovered legs. From her ankles upwards her legs had ballooned to twice their normal size and were violently discoloured. While she recovered in the A.I.M. Hostel in Alice, the children, once bathed and fed, seemed unaffected. As did most outback kids, they had been wearing overalls.

'Well, didden I bin learn im?' said Billy later when he heard about their survival. He was obviously a man of very few words.

Billy and Clara Walk into the Wind

When Glen Maggie had been sold, Claude took Billy and Clara to Woola Downs. Claude was taken ill some years afterwards and died in the Alice Springs Hospital, Mags and her family happened to be on the road droving cattle towards their own property, Ooratippra. As soon as Mags was able, she returned to Woola Downs to collect the couple, thinking to look after them in their declining years, but Billy and Clara had vanished. She was informed that they had last been seen heading in her direction, foot-walk, but despite an all-out intensive search they were never found. Their Aboriginal friends insisted that the old couple had walked into the wind. Another chapter was painfully closed.

Camel team heading past Mount Gillen to Alice Springs.

A photograph of Lasseter's grave taken by Bob Buck in 1932.

Mount Barker farewell. Ben 3rd from right. His mother, Liz, 7th from right in back row.

Double wedding in Colchester. Ben and Jane are front couple.

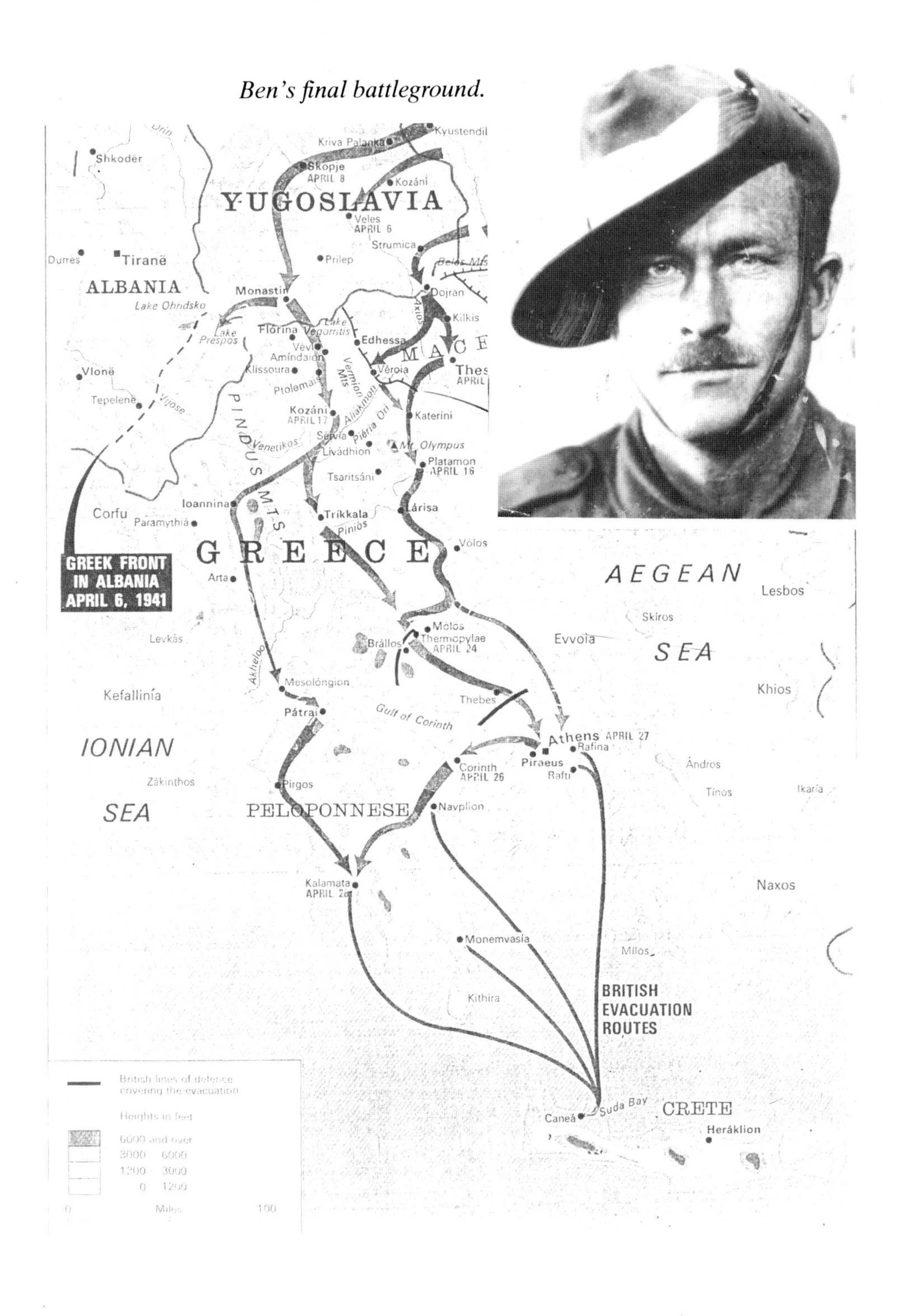

Ben's final battleground.

The last photograph of Ben.

NX 6237 Alan Low 2/3rd Field Regiment RAA laying poppies on SX403 Benjamin E. Nicker's grave at Phaleron War Cemetery, Athens. 2nd May during '50 Years of Peace Pilgrimage'. 1995.

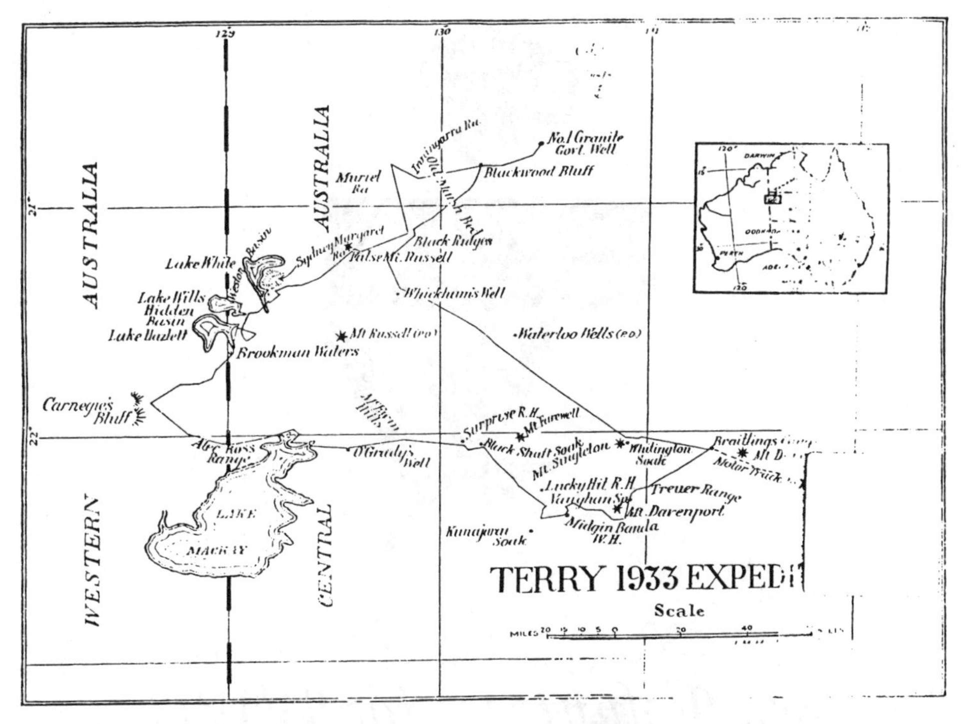

The 1933 expedition. Back row, Bob and Paddy.
L. to R. Stan O'Grady, Michael Terry and Ben Nicker.

7

Ben Volunteers for WWII

1939–1941

The 2nd/3rd Field Regiment

Ben was at Mount Doreen Station when the Second World War was declared. He wasted no time in heading south to enlist. He went to Adelaide and was directed to training camp at Woodside near Mount Barker in the Adelaide Hills. There they were joined by a unit from Rockingham in Western Australia who had reorganised at Northam before their transfer to Woodside. A real cross-section from every part of Australia came together with a single purpose but it must have been a very difficult task for their officers to weld such diverse personalities into a manageable fighting force.

Equipment was in short supply and much of their training was with imaginary guns. Frustration and personality clashes didn't make for an easy integration. It was a difficult time for men and officers alike.

In January 1940 they were moved to Ingleburn in New South Wales. Finally, they were supplied with equipment and began battle practice in surrounding areas. In a light-hearted moment, the 2nd/3rd commandeered the flag from the pole of the Log Cabin Hotel, Penrith, and took it with them overseas. It was returned in 1946 by sadly depleted survivors. They were the 2nd/3rd Field Regiment and were ready in May of 1940 to sail from Sydney for the Middle East on board the good 'Queen Mary.' Under heavy naval escort, they hove to near Fremantle, before settling down to their unexpectedly luxurious quarters on board and their long voyage to war. The 'borrowed' flag belonging to the Log Cabin Hotel fluttered from the 'Queen Mary' among official insignia and helped in a small way to alleviate the parting from families and lovers.

Expecting to reach the Middle East by way of Colombo, they were disconcerted when the 'Queen Mary' and her armada changed direction. Their first crossing the line was appropriately celebrated but their second was despondent. It took the appearance of Table Mountain to lift the gloom and shore-leave to colour the outlook. Let loose in Cape Town, they proved a difficult mob to muster back on board. It was said that 90% of the second/third were AWOL and that it took five days to retrieve them all. When the ship dropped anchor at Simonstown, their own MPs and the Africaan police were ready and waiting for any attempts to abscond and anyway, pocket money had already been either spent or handed over in fines.

Measles and mumps broke out, apparently picked up in Cape Town. It was a suffering rather than a jaunty regiment who were granted leave at Freetown, Sierra Leone. The grand 'Queen Mary' had become a floating hospital ward, but she ploughed on and into the *U-Boot* haunted seas and the lethargic patients were jolted into reality. They were escorted now by destroyers, an aircraft carrier, and the battleships 'Arc Royal' and the 'Hood.' The debris of a recent torpedo-strike littered the water surface around them and they were harried by submarine alerts. It was the war they had come for but they were still only passengers.

The Battle of Britain 1940

On the 16th of June 1940, they anchored at Graenock and were mesmerised by the gentle misted hues of Scotland as viewed from the deck, but it was with heavy hearts that they disembarked because Dunkirk was on every mind. While the 2nd/3rd approached, France fell. With the German advance to the English Channel, French and Allied troops were trapped on Dunkirk's shore. 338,000 troops were lifted off between May 29th and June 2nd 1940 by every available vessel, many of them civilian, and many thousands of lives were lost. Britain was struggling through her darkest hours.

The next day the Australians were moved by train to Tidsworth in Wiltshire, close by the beautiful city of Salisbury. The historic surrounds of their new camp should have been a source of wonder but the country was in trouble and these new troops were an embarrassment, having arrived unequipped. There wasn't any time to look around.

A major part of British transport and guns had been abandoned at Dunkirk. In partial solution, a quarter of the 2nd/3rd found themselves unhappily drafted to an infantry division and a great deal of shuffling about ensued among other companies. It was July before a half complement of eighteen pounders and howitzers gave the division the confidence to believe in themselves again.

They went into a period of intense specialised training and were finally issued with the regiment's new 25 pounder guns. Now their spirits were really lifted. The gunnery section was posted to the Royal Artillery School at Lark Hill near Stonehenge and they graduated with honours from that prestigious establishment.

The Daily Telegraph reported back to Sydney on the A.I.F. in Britain:

Leave for A.I.F. in Britain

> 'Leave has been renewed for sections of the A.I.F. This does not mean that headquarters believes the invasion of England is off. But it does show that the tension has eased slightly and that an immediate attack is considered less likely than it was a week ago.
>
> The A.I.F. hopes that an invasion will be attempted; they have no fear about the outcome.
>
> Diggers are angry about the arrogant behaviour of some of the German airmen brought down in England. A group of Queenslanders told me today 'There are a lot of things about the way the Germans fight that we don't like at all, especially their air raids. 'If any of them we get hold of start spitting on the ground and talking about 'English swine' they'll be sorry and we mean it. They had better behave themselves.'
>
> Stop Nazi Arrogance.
>
> An officer told me he thought there would be incidents if arrogant Nazis tried to do their stuff on the Diggers. 'In such circumstances I, for one would turn my back if the boys started anything,' he said. 'With German planes coming down all over England, the matter might arise at any time.'
>
> Australian gunners have profited enormously by lessons from Artillery experts.
>
> Yesterday I sat in an observation post while a Royal Artillery expert set problems in 'link shooting' to some young Australian officers. Among them were Captain F.A.Farmer of Toorak (Melbourne) and Lieutenant R.M.Davis of Western Australia.

> Yesterday the Australians accepted a challenge from a British battery. The British score was minus 4 points and the Australians' plus 25. The Australians achieved their success by not waiting to lay telephone lines but rushing up their guns to a ridge and firing over open sights.
>
> Among the gunners I saw yesterday was Bombardier Ben Nicker, well-known Northern Territory identity, mentioned in Michael Terry's book *Sand and Sun*. Nicker told me he had at last found his real mission in life.' From Ronald Monson, *Daily Telegraph* War Correspondent, London, Wednesday.

In a letter to Mags, dated August 12th 1940 Ben wrote:

> 'have been working hard for promotion and with some success as I am now a survey bombardier or in other words, the poor cow who does all the maths for the guns and wears two stripes. ....tobacco and smokes are a hell of a price and hard to come by so if you see anyone with more than he can smoke, tell him about us.'

The German invasion had begun and the 2nd/3rd were posted to a defensive position near Southampton. They were part of the Battle of Britain. Churchill could promise only 'blood, sweat, toil and tears'. While facing slow starvation under a *U-Boot* blockade, and the prospect of impending invasion, every heart and every pair of hands fought back. In October, mistakenly believing he had Britain cornered, Hitler turned his attention to the Balkans and the invasion of the Soviet Union.

Ben Falls in Love in Colchester

In October Ben's division was moved into the comparative luxury of Colchester Barracks accompanied by rumour of impending departure to the Middle East. Their equipment was shipped out but the troops were kept engaged in the defence of Colchester and in defence of their own attitude to authority. Their British compatriots found them impossibly lacking in respect for military tradition and that was unforgivable.

But Ben had no problems with the place or the people. He had met a lovely local army girl and quickly fallen in love. They courted for a month against a background of rubble and burning cities. A special licence, an expensive manoeuvre, enabled them to marry in ten days instead of waiting the usual three weeks, but time for them was short. They were deeply in love and hanging over them was the knowledge that the regiment would be posted overseas at a moment's notice.

There wasn't a honeymoon and having nowhere to live they moved in with Jane's mother. His commanding officer gave him a sleeping-out pass and they were grateful to make do with that and their precious time together. Jane was up early every morning to cook his breakfast before he reported for duty at six o'clock and had, despite stringent rationing, a hot meal waiting for him at night.

Winter set in and the Australians found it bitter beyond expectations. They slept in their uniforms and overcoats and grumbled while they waited to move out.

It was the end of the year before orders came and while for most it was a welcome relief, there were a few like Ben whose hearts had betrayed them. The excitement and ideals which had brought them to the brink of war had lost their impetus and foundered in love and marriage in England even under such desperate conditions.

He and Jane had enjoyed six weeks together before she saw him off on the troop train on a cold, wet, wretched morning. Jane wrote about their farewell:

> '......our parting was almost unbearable. Early morning, cold wet and thoroughly Hell Let Loose. Everyone crying seeing them off........he did not want to leave me. It broke my heart, I felt the bottom had dropped out of my world........'

Service in the Middle East

On December 18th 1940, the unit boarded a ship for their original destination, the Middle East. They again rounded the Cape and sailed through the Suez Canal then to Alexandria where they were loaded aboard troop-trains to Ikingi Maryut and the regiment was split into two divisions.

In one of his letters home Ben wrote:

> '.... and speaking of voyages, this regiment seems to be on a Cook's Tour or else like Van der Bekken, someone in it has doomed us to travel for ever and not catch up with the war, still, this may take us somewhat nearer to it. Have been in numerous camps in a few countries, among them England as you know by now, Scotland in summer and again when the snow was on the highlands, the mountains of Wales and over England, and of all the scenery that of the Cotswolds is the best. Difficult to find space to write but somehow I manage, ca ne fait rien, I may manage to keep it up.....'

> '.......four piasters a time keeps my literary efforts within bounds, not that I was ever a Milton. As a campsite the present one compares very well with Oodnadatta except that there is even less growth.............I often think how easily things could have been

> different, for instance, if Madam Hitler had only read Dr Stopes. Small things but what a mess. Still if it had not have happened I would not have met my Jane and that would have been my loss....and may be her gain...kismet.............time and paper slip by so I will say goodbye for a time and if after a decent interval you don't hear from me, please write the usual rubbish to my wife..........'

While RHQ and the 5th Battery went into action in the Western Desert as far as Benghazi, the 6th Battery were employed around Burg-el-Arab, El-Alamein and Cairo. This was a different war. British Commonwealth forces were engaged in the liberation of Abyssinia from Italy and the defeat of Mussolini's troops in Libya.

In another letter Ben wrote:

> '................the usual eastern town, flat-roofed white houses running right down to the sea. Some of the boys have gone down for a swim, but as the day is cold, I, naturally, restrain any foolish instinct towards cleanliness and remain lousy, dirty and comfortable.........How goes everything at home? One wonders what's happening in the scenes of one's civil indiscretions and if you know and I am sure you do, all the scandals, don't be so mean about passing some along. I met Dutchie for a few minutes only and as the canteen had closed we weren't able to do the occasion justice but hopefully some other day, some other canteen.
>
> And now, old pal, I guess I had better close, at least for a time, if otherwise, vale.'

The Campaign in Greece

Re-united by the end of March, the entire unit embarked for Greece, where German invasion was imminent. The Greeks had repulsed Mussolini's troops in the previous year but now Hitler's forces had ploughed through the Balkans, to Yugoslavia and were hovering over Greece.

Landed at the Port of Piraeus on April 2nd 1941, the division enjoyed a brief respite until Hitler made his move. They were immediately ordered to Servia through extraordinary mountains by way of Athens, Alanti, Thermopylae, Lamia, Volos, Larissa and Elassona. For these Australians the route was staggering. They had never seen, let alone driven through such heights. Their goal was Servia Pass and it was here that the 2nd/3rd fired the first shots of the Allied Forces in Greece. Conditions were fierce. They were 3,000 feet above sea level and it was snowing. Ben wrote more letters, daily, if possible:

> '....you know, things are never so bad that they couldn't be worse although there was a time up at the line when fourteen German bombers paid us a call and stayed to empty their guns at us, I thought it couldn't be any tougher but the damage wasn't too bad. Anyhow that phase of the war is over for us or at least it is pro tem for now we are resting, for what, quien sabe? Well, old pal, we went through a fair bit of the Lybian Campaign including Tobruk, Derna, Bengazi and some other towns whose names I have forgotten. This must be one of the fastest campaigns ever fought and, I hope, the turning point of the war.
>
> I had leave in Alexandria and did the place as a tourist and then as a soldier. The first goes where he is taken and the second where he darn well pleases.
>
> ....the bombing back in England doesn't sound to be any easier and I worry about Jane. At least she won't be afraid of the bush when this is all over.'

The meagre, snow-thralled road they followed through Greece, he knew to be the battle-path of history. The same route had been well-worn and blood-soaked by many nations' warriors, an ongoing four thousand-year saga of which he was himself now part. They had reached the slopes of Mount Olympus, the highest point in Greece.

It was once believed the earth was round and flat and that Mount Olympus was the central point, the home of the gods, hidden from mortal view by a permanent wreath of cloud. Goddesses known as the Seasons governed the gateway to the heights from where Jupiter ruled the earth and his dynasty of gods with that devilish weapon, thunder. Jupiter with his thunder bolts had never been a favourite character of Ben's.

Ben's Death in Athens

The German attack came on April 11th. The 6th Battery stayed in action while the 5th withdrew to a higher position from where they were able to cover the fall-back of the 6th. In the early morning darkness the regiment moved back to Servia and positioned themselves south of the Aliakmon River. Here they suffered their most intense enemy dive-bombing from a flight of twenty two Stukas.

NX 6237 Gunner Alan R. Low remembered:

> 'At Servia Pass in Central Greece in April 1941 on our retreat from the Greece/Yugoslavia border we came under heavy German aircraft attacks on our gun emplacements and had to prepare for a quick withdrawal south. The valley of the Pass which we had to

cross to climb the escarpment on the other side was coming under heavy German artillery fire. Our vehicles were being waved through two at a time by Major Dicky Bale WX27 between German salvos. I was driving a 3-ton ammunition truck which had several boxes of 5th.battery kitchen utensils piled on top of the ammunition with some of the cooking staff. When I was waved off, I let her rip to get up as much momentum to hurry up the steep narrow winding road on the escarpment on the other side of thc valley.

About half way up the escarpment, and by this time I was in low gear, we again came under heavy German artillery fire with shells bursting around us. One shell burst very close to the passenger side door and shrapnel hit the mudguard, windscreen, tore through the truck canvas canopy hitting boxes containing the kitchen utensils. Fortunately the three Gunners on the back namely John (Jeep) Butler NX11053, Colin (Lubra) Murray DX84 and Cecil (Spud) Bailey TX? were not injured except a near miss for Gunner Bailey when a shrapnel put a dent in his steel helmet.

Unfortunately we didn't miss out in the driver's cabin. Both Gunner Cornish (Con) Dolan DX9 and myself received superficial cuts from the shattered windscreen. Bombardier Ben Nicker SX403 was sitting on the passenger seat next to Gunner Dolan sitting on the battery box in the middle. A piece of shrapnel had pierced the passenger door then through the seat into Bombardier Ben Nicker's buttock. Ben let out a yell 'I've been hit.' The shock caused him to become pale in the face and turned on his side to try and ease the pain with Gunner Dolan holding his head and doing his best to comfort Ben.

While all this was happening I kept driving to get out of the German artillery range of fire for we had been told under no circumstances were we to hold up the convoy as they could not pass on the narrow mountain road and would be sitting ducks for the German artillery. As I was in low gear it seemed an eternity before we passed out of range of the shelling and reached the top of the escarpment. About a mile on where the road widened I pulled to the side to see what could be done for Ben and inspected the truck's damage.

We lifted Ben out of the truck and laid him on his side on a ground sheet. Now the initial shock had passed and more colour came into Ben's face but this was only natural, for Ben was one tough fella. I always carried several bottles of Cognac in one of the empty cartridge cases purely for medicinal purposes. We gave Ben a liberal swig which brought more colour to his face and I then told him to hold his breath while I poured some on his wound to help clean it. Surprisingly the shrapnel which was about two inches in diameter and very ragged had penetrated sideways in the buttock and seemed to penetrate about three quarters of an inch. There was hardly any bleeding for the shrapnel was red hot when it lodged and seemed to sear the wound.

We were wondering what was the best thing to do when Lieut. Trevor West NX374 with his driver pulled up in his utility and wanted to know why we were stopped. He realised our problem when he saw Ben lying on the ground sheet and the damage to the truck. It was obvious Ben had to get to a casualty clearing station as quickly as possible so we placed Ben in the back of the utility under the canvas canopy and Lieut.West sped off to the nearest casualty clearing station.

After making a thorough inspection of the truck we found it was mechanically and structurally sound so we headed off after Lieut. West. Some ten miles further on we came across the casualty clearing station which had been obviously rushed into service with three tents for operations and recovery. The patients were lying on stretchers and ground sheets, many with serious wounds, stomach, head and limbs.

Several ambulances and trucks were on stand-by for an immediate withdrawal. We found Ben lying on his side on a ground sheet and he had he not been attended to. This was understandable for there were many around him in a hell of a mess.

Con Dolan and 'Lubra' Murray who were particular friends of Ben were with me. Ben was trying to make light of his wound and said that when they got the shrapnel out and stitched him up he would get back to the Regiment as quickly as possible. I left Ben with two bottles of Cognac as we had to leave, for the tail end of our regiment's convoy was passing on the road.

Nine days later we received the tragic news Ben had passed on in the military hospital in Athens from gangrene.'

The unit were forced to again withdraw this time to Ellasona where their guns ran so hot that paint peeled from their barrels. Under heavy shelling the regiment were pushed back to Port Raffina and there shipped out under cover of night, by barge and warship to Crete. In Crete they were re-armed and positioned to defend the island's aerodromes and deployed to the defence of Suda Bay. The 2nd/3rd lost two-thirds of its men and twelve of its fifteen officers. The survivors who managed to escape over the mountains were taken off by the navy or managed to struggle through somehow to Alexandria.

A fragmented country lay behind them. 300,000 people died in Greece in that winter's cold and famine. 60,000 Jews were rounded up by the Germans and despatched to extermination in Poland from which horror a mere 6,000 escaped.

Australian troops in Crete before the German attack were estimated to number 6,486. A large number were made prisoners of war. Commenting in June 1941 on Australian losses in Crete and Greece, Dr C.E.W. Bean, Australian Official War Historian, said that the battles of Greece and Crete together might have cost Australia almost 2,000 human casualties.

Few of Jane's letters ever reached Ben and for months after his death his letters to her kept arriving, despite heavy censoring, full of love and promises he was unable to keep.

Recorded at Phaleron War Cemetery, Athens.

Nicker, Bdr., Benjamin Esmond, S.X.403,AIF 2/3rd Field Regt., Royal Australian Artillery. 19 April,1941. Aged 33. Son of Samuel Foreman Nicker and Elizabeth Nicker of Alice Springs, Northern Territory, Australia. Plot 3. Row B. Number 5.

Ben Nicker: Outstanding Soldier, Intrepid Bushman

'In Greece, Ben's unit was pinned down by German guns, which could not be pinpointed. Ben volunteered to go out by himself and find them, but his commanding officer, Major McNamara, refused to risk him.

Ben went anyway. Three days later he was back with a detailed account of the positions of the German guns. Bombardier Nicker should have got the D.C.M. for that exploit, but as it was done against orders, he was not decorated........

Fred Colson was right when he said that Ben Nicker was one of our finest bushmen. I knew him for twelve years......We must have put up a world record, no row, not even a cross word during our time in the wilds. Ben's nature contributed greatly towards this. He had an exceptional sense of humour and would laugh himself into tears over a remark or an incident. Further, having been born and bred among blacks, he was the best tracker I have met.

On several occasions I have seen him correct blacks about points in tracking. He often talked about outstanding camel men but would never acknowledge that his knowledge was exceptional. He was indeed a finished camel man of whom there are only a handful in Australia. In Ben's death I have lost a great friend, with whom as soon as this war ends I had meant to go bush. I well remember, some years ago when a trip was mooted, which my health frustrated, I wired Ben to come. He replied, 'Yours to the last pituri cigar.' Stan O'Grady too will miss him as much as I do. We were both more than ordinarily fond of our mate with whom we nearly perished twice, once by thirst and once by blacks.

Out west in a place which we found and named 'Hidden Basin', in 1933, on the border of Western Australia, we named a creek Nicker Creek after Ben. That name is cut into the trunk of a large gum tree. Ben Nicker devoted his life to the bush beyond fences. He has given his life to keep the bush free for you and me. I grieve for his wife, and with the bushmen of Central Australia I mourn the passing of a very fine fellow.'

(1941) Michael Terry.

*

'The sun shone warmly from a Grecian sky. The corn was a rich gold as it gently waved toward Olympus, and big red poppies lifted their heads to the heavens in a pine-scented wind. Ben and I laughed, we had to laugh for we were warm with wine and temporarily in love with the war. It was on the coast road going up. We sat under the tall pines looking across the inland sea of the Negran, to the high snow-draped mountains. We all laughed for our pockets were full of drachma.

Ben often told us stories of his youth; days spent wandering around the Dead Heart of Australia, and crossing the notorious Simpson Desert. He knew a lot.

It was always 'Ask Ben.' He could pull any gun to pieces and put it back again. I can hear him now, 'Wish I could get a few letters from my wife..... they take so damned long to come from England'. We all had a great time the night he married a pretty girl from Colchester. It must have been a hard break to leave England. Later the guns opened up in the Battle of the Bog. All night they spat flame and death at the Hun. We wondered about their Air Force. So did Ben, for he was our ack-ack machine gunner and was ready for action. When our Major told us the rain and snow were keeping his planes grounded, Ben cursed. But they came, squadrons of them over Servia Pass. You could not buy a pick or shovel for love nor money.

Ben had his pit between the command post and the ack-ack troop out in the open on a slight rise. Whiskey, his offsider, had his ammo stacked everywhere. First came a Henschel spotting for German artillery and Stukas. The big guns held their fire; so did Ben. Yet the spotter must have seen the guns for a score of Stukas circled their positions and peeled off one by one.

We watched Ben from our slit trenches, he gave them all he had. It gave us strength to see Whiskey dash across open ground for more ammo. All that afternoon they attacked and Ben kept at it, blazing away, he was black with dirt and smoke, his hands were scorched through changing the red-hot barrel, but he didn't leave his guns. Not far from him a water-cart received a direct hit with a 100 pounder and half filled his pit, only dusk brought peace and a spell for Ben. Dawn came and with it Goering's famous yellow-nosed fighters. Bullets spattered like raindrops around Ben for what seemed like hours. He helped to drive the fighters off ... then came the Hun through the pass ... we were the rear guards so the guns had to go out fast. Long range guns were shelling our positions and the road was out. Ben was in the truck ahead of me when shrapnel struck him.

Months afterwards back in camp near Tel Aviv, I was sorting the mail and I placed aside a special heap of letters from England that Ben would never read. We learned in Crete that he had lain for seven days in the same field dressing in a Daphne hospital and died with gangrene.

Perhaps Ben wasn't a real hero. Perhaps he never died a hero's death, but he was one in our eyes in those never-to-be-forgotten days. Wherever he sleeps, I pray the snow falls softly there and that the poppies are gently fanned by those same winds scented with pine.'

'Ben was a hero in their eyes. B.D.R. Ben Nicker' by SX137 Gnr L.W. Sunman, *Ack-Emma Foofs*, printed during World War Two, date unknown.

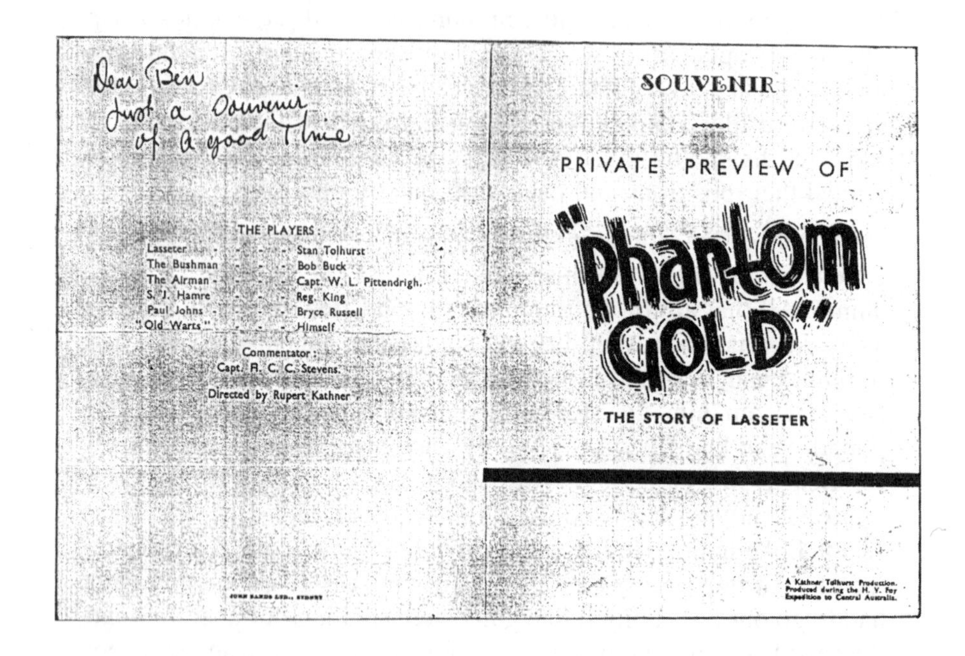

Dear Ben
Just a Souvenir
of a good Time

THE PLAYERS:

Lasseter	Stan Tolhurst
The Bushman	Bob Buck
The Airman	Capt. W. L. Pittendrigh.
S. J. Hamre	Reg. King
Paul Johns	Bryce Russell
"Old Warts"	Himself

Commentator:
Capt. R. C. C. Stevens.

Directed by Rupert Kathner

JOHN SANDS LTD., SYDNEY

SOUVENIR

PRIVATE PREVIEW OF

"Phantom GOLD"

THE STORY OF LASSETER

A Kathner Tolhurst Production.
Produced during the H. V. Foy
Expedition to Central Australia.

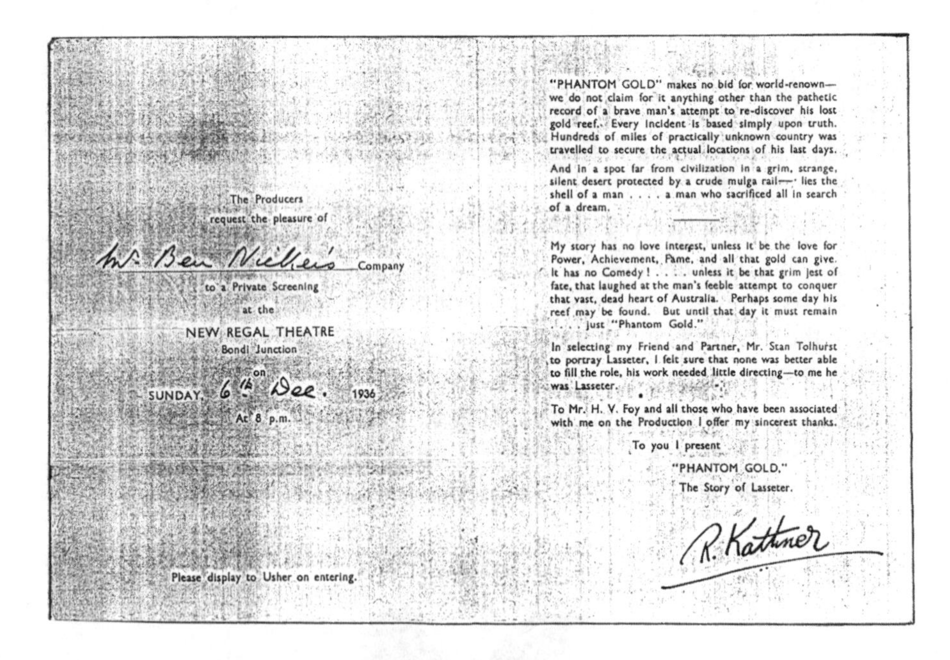

The Producers
request the pleasure of

Mr. Ben Nickers Company

to a Private Screening
at the
NEW REGAL THEATRE
Bondi Junction
on
SUNDAY, 6th Dec. 1936
At 8 p.m.

Please display to Usher on entering.

"PHANTOM GOLD" makes no bid for world-renown—we do not claim for it anything other than the pathetic record of a brave man's attempt to re-discover his lost gold reef. Every incident is based simply upon truth. Hundreds of miles of practically unknown country was travelled to secure the actual locations of his last days.

And in a spot far from civilization in a grim, strange, silent desert protected by a crude mulga rail— lies the shell of a man . . . . a man who sacrificed all in search of a dream.

My story has no love interest, unless it be the love for Power, Achievement, Fame, and all that gold can give. It has no Comedy! . . . . unless it be that grim jest of fate, that laughed at the man's feeble attempt to conquer that vast, dead heart of Australia. Perhaps some day his reef may be found. But until that day it must remain . . . . just "Phantom Gold."

In selecting my Friend and Partner, Mr. Stan Tolhurst to portray Lasseter, I felt sure that none was better able to fill the role, his work needed little directing—to me he was Lasseter.

To Mr. H. V. Foy and all those who have been associated with me on the Production I offer my sincerest thanks.

To you I present
"PHANTOM GOLD."
The Story of Lasseter.

R. Kathner

Phantom gold. Ben's invitation to the film about Lasseter produced during the H.V. Foy expedition and reviewed in 1936.

Epilogue

'I was pleased last week to meet Jack Lowe, just back from one of his trips to the Northern Territory. He showed me a letter he had from abroad from Lieutenant Harold (Tiger) Brennan, well known in Alice Springs and the Wauchope mines country. 'Poor old Ben Nicker met his end,' he writes; 'but if everyone did in the whole war what he did in the last 15 months, the war would have been over long ago. He brought down three German dive bombers in quick succession before a bomb got him.' '

The Advertiser, 'Vox', 'Boys of The Bush'

*

'Of all the interesting characters occasionally met with by far the most interesting and unusual was Ben Nicker. I have never met anyone quite like Ben before or since. He had a better education than was usual to find in those days........Ben always wanted to be a soldier which is unusual for a man of his type, they can't take military discipline as a rule. When war broke out in 1939 Ben was one of the first to enlist in the second A.I.F. He was killed in the evacuation of Greece in 1941. He was a marvellous rifle shot and I was told by someone who was with him in Greece that he shot down a German plane with the ordinary .303 rifle by shooting the pilot in the cockpit as he flew over.'

Bryan Bowman, *History of Central Australia. 1930–1980.*
(A desk-top publication)

*

A great niece of Ben's is today a cameleer, trekking the outback. Whenever and whatever problems arise with her team, she asks herself, 'What would Ben have done?'

'Sometimes,' she says, 'I'm certain he's travelling with me.'

A great nephew is Ricky Hall, the cameleer who received national notoriety when he was charged with driving a camel wagon while under the influence of alcohol. He was acquitted.

*

When Mags received the Order of Australia Medal in 1986 for her services to Central Australia, she said 'I accept this on behalf of my family and their contribution to the outback.'

*

Out in the deserts the name 'Benninik' comes up conversationally in Aboriginal dialects more than fifty years after his death. Travelling in four-wheel drives equipped with two-way radios, modern adventurers frequently come across his carved initials.

When next the question is asked, 'Who was Ben Nicker?' I hope some of the answers will be found in these pages.

References

The Last Explorer by Michael Terry, FRGS, FRGSA. Compiled by Charlotte Barnard. Printed by Australian National University Press in association with North Australia Research Unit (Darwin), 1987.

The Man from Oodnadatta by R.B. Plowman, Angus and Robertson, 1933.

This book and others by R.B. Plowman abridged and republished by his daughter, Jean Whitla. Shoestring Press, Wangaratta, Victoria. First edition 1992. *The Man From Oodnadatta.* 'Stories by the First Patrol Padre to the Australian Inland Mission.' 1912–1917.

A History of Central Australia. 1930–1980 by Bryan Bowman. A desk-top publication. Undated.